Liquid Assets

THE CRITICAL NEED TO SAFEGUARD
FRESHWATER ECOSYSTEMS

SANDRA POSTEL

Lisa Mastny, *Editor*

WORLDWATCH PAPER 170

July 2005

THE WORLDWATCH INSTITUTE is an independent research organization that works for an environmentally sustainable and socially just society, in which the needs of all people are met without threatening the health of the natural environment or the well-being of future generations. By providing compelling, accessible, and fact-based analysis of critical global issues, Worldwatch informs people around the world about the complex interactions among people, nature, and economies. Worldwatch focuses on the underlying causes of and practical solutions to the world's problems, in order to inspire people to demand new policies, investment patterns, and lifestyle choices.

FINANCIAL SUPPORT for the Institute is provided by ACORE (American Council on Renewable Energy), the Aria Foundation, the Blue Moon Fund, the Fanwood Foundation West, the German Government, Goldman Environmental Prize/Richard & Rhoda Goldman Fund, the W. K. Kellogg Foundation, The Frances Lear Foundation, the Steven C. Leuthold Family Foundation, the Massachusetts Technology Collaborative, the Merck Family Fund, the Norwegian Royal Ministry of Foreign Affairs, The Overbrook Foundation, the V. Kann Rasmussen Foundation, the Rockefeller Brothers Fund, The Shared Earth Foundation, The Shenandoah Foundation, the Tides Foundation, the United Nations Population Fund, the Wallace Global Fund, the Johanette Wallerstein Foundation, and the Winslow Foundation. The Institute also receives financial support from many individual donors who share our commitment to a more sustainable society.

THE WORLDWATCH PAPERS provide in-depth, quantitative, and qualitative analysis of the major issues affecting prospects for a sustainable society. The Papers are written by members of the Worldwatch Institute research staff or outside specialists and are reviewed by experts unaffiliated with Worldwatch. They have been used as concise and authoritative references by governments, nongovernmental organizations, and educational institutions worldwide. For a partial list of available Worldwatch Papers, go online to www.worldwatch.org/pubs/paper.

The views expressed are those of the authors and do not necessarily represent those of the Worldwatch Institute; of its directors, officers, or staff; or of its funding organizations.

Contents

Acknowledgments: I am grateful to Brian Nicholson for his valuable research assistance on this project. I extend thanks to water experts Ellen Douglas, Carmen Revenga, Amy Vickers, and Charles Vörösmarty, as well as agricultural specialist Brian Halweil, for their helpful comments on a draft of the manuscript. I also thank Chris Flavin, Worldwatch's president, for the opportunity to write this paper under the Institute's banner, Lisa Mastny for her judicious and careful editing, Lyle Rosbotham for transforming the manuscript into these handsome pages, and Patricia Shyne, Darcey Rakestraw, and Courtney Berner for their skillful help with marketing and outreach.

Sandra Postel is Director of the Global Water Policy Project in Amherst, Massachusetts, Visiting Senior Lecturer in Environmental Studies at Mount Holyoke College, and a senior fellow with Worldwatch. She is author of *Last Oasis* (1992), the basis for a 1997 PBS documentary, and of *Pillar of Sand: Can the Irrigation Miracle Last?* (1999). Her most recent book, co-authored with Brian Richter, is *Rivers for Life: Managing Water for People and Nature* (2003). Sandra's article "Troubled Waters" appears in the 2001 edition of *Best American Science and Nature Writing*. In 2002, she was named one of the "Scientific American 50" by *Scientific American* magazine, an award recognizing contributions to science and technology.

SUMMARY

By taking advantage of the work that healthy watersheds and freshwater ecosystems perform naturally, cities and rural areas can purify drinking water, alleviate hunger, mitigate flood damages, and meet other societal goals at a fraction of the cost of conventional technological alternatives. But because commercial markets rarely put a price on these "ecosystem services," and because governments are failing to protect them, they are being lost at a rapid rate.

Rivers, lakes, wetlands, and other freshwater ecosystems provide a host of benefits to society. In addition to supplying water and fish, they store seasonal floodwaters, helping to lessen flood damages. They recharge groundwater supplies, which can ensure that water is available during dry spells. They filter pollutants and purify drinking water. And they provide the diverse habitats that support the myriad species performing much of this ecological work.

The water strategies of the twentieth century worked largely against nature, not in partnership with it. Dominated by large dams, levees, river diversions, and other big engineering projects, these strategies helped provide much of the world with drinking water, food, electricity, and flood control. But they also disrupted the functioning of aquatic ecosystems on a vast scale. Today, dams and reservoirs intercept about 35 percent of river flows as they head toward the sea, up from 5 percent in 1950. Many overtapped rivers no longer reach the sea for extended periods of time, ruining deltaic fisheries and degrading coastal zones. Meanwhile, the loss of wetlands,

floodplains, and forested watersheds is increasing the economic and human toll of floods and other natural disasters,
a toll almost certain to increase as global warming intensifies
the hydrological cycle.

Meeting new water needs requires a different approach.
Fortunately, forward-thinking cities, villages, and farming
regions around the world are demonstrating that drinking
water, food security, and flood control needs can be met in ways
that employ, rather than destroy, ecosystem services. Many are
saving their residents money by doing so, while creating
healthier environments at the same time.

A variety of municipalities are realizing, for example, that
healthy watersheds are nature's water factories, and that it
pays to protect them. More than half a dozen U.S. cities have
avoided the construction of expensive filtration plants by protecting their watersheds. Working in partnership with towns,
businesses, and community groups in the Catskills-Delaware
watershed, New York City is investing $1.5 billion in watershed
measures over 10 years to avoid a filtration plant estimated to
cost $6 billion to build and $300 million per year to operate.

Bogotá, Colombia, and Boston, Massachusetts, have coupled watershed protection with effective conservation efforts,
reducing capital expenditures and safeguarding ecosystem
services. Bogotá's conservation success has delayed the need
to construct new water supply facilities for at least 20 years.
Water use in the greater Boston area hit a 50-year low in 2004,
following an aggressive conservation program begun in the late
1980s that has indefinitely postponed construction of a diversion from the Connecticut River and saved residents more than
$500 million in capital expenditures alone.

Strategies that value and protect watersheds, wetlands, and
floodplains are also critical to reducing hunger, which now saps
the health and energy of 852 million people, most of whom
live in poor farming regions. Rainwater harvesting methods
coupled with affordable small-plot irrigation technologies are
enabling poor farmers to boost their crop production by using
local water supplies more effectively. Better management of
soils, water, and nutrients has quadrupled irrigated rice yields

in parts of Madagascar, while also saving water. And researchers studying an extensive floodplain in northeastern Nigeria found that the net economic benefits of direct use of the floodplain—for agriculture, fuelwood, and fishing—exceeded by 60-fold those of an upstream irrigation project that would destroy much of the floodplain.

Healthy ecosystems provide valuable insurance against catastrophic losses from flooding and other natural disasters as well. Nearly 5,000 Haitians lost their lives and tens of thousands lost their homes during tropical storms in 2004. Although tagged as natural disasters, these tragedies were exacerbated by the clearing of trees in the Haitian highlands, which left sloping lands susceptible to rapid flood runoff and massive mudslides. The same storms that devastated Haiti caused far less damage in neighboring Puerto Rico, where highland watersheds are mostly forested.

For the same reason people buy home insurance and life insurance—to avoid catastrophic losses—societies need to "buy" disaster insurance by investing in the protection of watersheds, floodplains, and wetlands. Global warming and its anticipated effects on the hydrological cycle will make the robustness and resilience of nature's way of mitigating disasters all the more important, as tropical storms, spring flooding, and seasonal droughts increase in frequency and intensity.

Governments need to overhaul water policies and practices in a way that will protect freshwater ecosystems and their valuable services. High priorities include requiring drinking water suppliers to invest in watershed protection; inventorying and setting ecological goals for the health of rivers, lakes, and other freshwater ecosystems; and establishing caps on the degree to which human activities modify river flows, deplete groundwater, and degrade watersheds. Combined with more effective water pricing, these caps will drive up water productivity—the unit value of water extracted from nature—and help to meet human needs while safeguarding nature's vital freshwater ecosystems.

Introduction

In 2005, engineers began constructing a dike 6 meters high and 13 kilometers long to try to prevent the northern part of Central Asia's Aral Sea from draining completely. Once the world's fourth largest lake, the Aral Sea is now a poster child of aquatic ruin. A half-century after Soviet officials decided that the two rivers feeding the lake would be more valuable for irrigating cotton, the Aral has lost 80 percent of its water. It has split into two parts: a small lake in the north and a larger one in the south. Most of the fish—and 60,000 fishing jobs—have disappeared. Winds crossing the exposed seabed pick up millions of tons of salt and dust laden with pesticide residues, poisoning the air and land. The 3 million people in the "disaster zone" suffer from high rates of cancers, respiratory ailments, anemia, and other illnesses. Thousands have fled the area.*[1]

The dike is a desperate attempt to arrest this cascade of unintended consequences. Officials hope to raise the Small Aral by 3 meters, cover some 1,000 square kilometers of desiccated seabed with water, increase rainfall, and lessen the dust storms. But the dike is another gamble with nature. Although it is likely to revive the small sea, it may further shrink the big one and compound the problems around it.

Although extreme, the Aral Sea disaster is not unique. Human actions have altered rivers, lakes, and other freshwa-

---

*Units of measure throughout this paper are metric unless common usage dictates otherwise. Endnotes are grouped by section and begin on page 61.

ter ecosystems in numerous ways and places. And despite the warnings sounded by the Aral Sea's demise, officials keep rolling out ecologically destructive new water schemes as if blind to the risks and ramifications.

Earth's hydrological cycle—the sun-powered movement of water between the sea, air, and land—is an irreplaceable asset that human actions are now disrupting in dangerous ways. Although vast amounts of water reside in oceans, glaciers, lakes, and deep aquifers, only a tiny share of Earth's water— less than one-hundredth of 1 percent—is both fresh and renewed by the hydrological cycle. That precious supply of pre-cipitation—some 110,000 cubic kilometers per year—is what sustains most terrestrial life.[2] (See Figure 1.)

Like any valuable asset, the global water cycle delivers a steady stream of benefits to society. Rivers, lakes, wetlands, aquifers, and other freshwater ecosystems work in concert with forests, grasslands, and other landscapes to provide goods and services of enormous value to human society, from miti-gating floods to recharging groundwater. (See Sidebar 1, page 12.) As the people of the Aral Sea region know firsthand, the nature and value of those services can remain grossly under-appreciated until they are gone.

From hunter-gatherers to the most advanced irrigation-based communities, human societies have always depended on freshwater ecosystems for food, water, and livelihoods. It is no coincidence that the great early civilizations sprung up and flourished alongside rivers. The ancient Egyptians thrived for several thousand years on the Nile's annual flood, which delivered water and nutrients to their fields and carried off harmful salts that had accumulated in the soil.[3] Yet history is equally clear that dependence on nature's services comes with a responsibility to protect them. An unsettling number of societies have effectively collapsed from what author Jared Dia-mond terms "self-inflicted ecological suicides," the destruction of the ecosystems upon which those societies depended.[4] Likely members of this group include the Sumerians of ancient Mesopotamia, the Harappan of the Indus River valley, and the Anasazi, Hohokam, and Maya of the Americas.[5]

FIGURE 1

The Global Hydrological Cycle

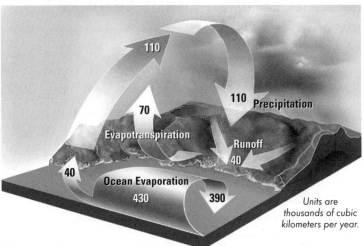

Note:
Precipitation over land (110,000 cubic kilometers) constitutes the total renewable water supply available to support non-marine plant and animal life and all terrestrial and fresh-water ecosystems. It divides into two major parts: evapotranspiration and runoff.
Evapotranspiration is water returned to the atmosphere by evaporation from land, water, or other surfaces or from transpiration by plants. Evapotranspiration from land-based eco-systems (70,000 cubic kilometers) is the renewable water supply for forests, grasslands, rainfed croplands, and all other non-irrigated vegetation.
Runoff is water that flows in rivers, streams, or underground aquifers back toward the sea. This 40,000 cubic kilometers is the water supply for irrigated agriculture, industries, and cities, as well as for all species dependent on rivers, streams, and groundwater systems.
Source: See Endnote 2 for this section.

Today we are apt to think that our globalized and tech-nologically sophisticated world is immune to such societal col-lapse. But there is no side-stepping human dependence on the water cycle. More than 99 percent of the world's irrigation, industrial, and household water supplies come directly from rivers, lakes, and aquifers.* Wetlands and river floodplains protect people from floods, provide spawning habitat for fish, recharge groundwater supplies, renew soil fertility, and purify water of contaminants. In the Mekong River basin of south-east Asia, for example, more than 50 million people depend

---

* Less than one-half of 1 percent of world water use comes from desalination.

SIDEBAR 1

Life-Support Services Provided by Rivers, Wetlands, Floodplains, and Other Freshwater Ecosystems

- Water supplies for irrigation, industries, cities, and homes
- Fish, waterfowl, mussels, and other foods for people and wildlife
- Water purification and filtration of pollutants
- Flood mitigation
- Drought mitigation
- Groundwater recharge
- Water storage
- Wildlife habitat and nursery grounds
- Soil fertility maintenance
- Nutrient delivery to deltas and estuaries
- Delivery of freshwater flows to maintain estuarine salinity balances
- Aesthetic, cultural, and spiritual values
- Recreational opportunities
- Conservation of biodiversity, which preserves resilience and options for the future

upon fish for their nutrition and livelihoods, and 90 percent of those fish spawn in the fields and forests of the river's floodplain.[6] Healthy river systems are also vital to life in lakes, estuaries, and many coastal marine environments. Their flows deliver the nutrients and maintain the salinity balances so critical to many fisheries, from the prized blue crabs and oysters of Florida's Apalachicola Bay to the lost fishes of the Aral Sea.

Scientists are working to determine more precisely how the plants, animals, and environments in which they live provide these services. For their part, economists are attempting to place monetary values on these services so that decisionmakers can better take them into account. In the meantime, however, ecosystem disruptions continue at an accelerating pace as growing populations and economies place new demands on land and water.

Society faces a serious conundrum: competition for water is not only intensifying within and between countries, but also between people and the ecosystems on which they depend. To support the diets of the additional 1.7 billion people expected to join the human population by 2030 at today's average dietary water consumption (the rainfall and irrigation water consumed in producing the average diet) would require 2,040 cubic kilometers of water per year, a volume equal to the annual flow of 24 Nile Rivers.[7] What is the ecological cost of appropriating this much additional water from natural systems? Using more rainfall for food production typically means clearing more forests to make way for crops, while using more irrigation water means further straining rivers, lakes, and aquifers. More importantly, are there ways to satisfy human needs for water and food without sacrificing a dangerously high proportion of ecosystem services?

Governments implicitly committed themselves to strive to answer this question when they unanimously endorsed the United Nations' Millennium Development Goals (MDGs) in September 2000.[8] The MDGs include specific targets of halving by 2015 the proportion of people living on less than $1 a day, suffering from hunger, and lacking access to affordable safe drinking water.* As if meeting these targets will not be difficult enough, another one calls on the world community to do so while ending the unsustainable exploitation of natural resources.

To achieve these objectives and avoid more Aral Sea disasters, political leaders will need to turn water management on its head. Protecting the health and functioning of freshwater ecosystems needs to be a top priority, not the bottom rung on the ladder. Long the domain of hydrologists and civil engineers, water management needs to include foresters, soil scientists, ecologists, anthropologists, and others who understand the wide range of benefits that ecosystems provide and how nature provides them. Mike Dombeck, former chief of the U.S. Forest Service, reframed the role of the national forests when he wrote in the *New York Times* that "the focus should be on how to let our forests do their job of producing high-quality

---

* Unless otherwise noted, all dollar figures are in U.S. dollars.

water. Given our water supply problems, this should be the highest priority of forest management."[9]

A small but critical mass of cities, villages, and farming regions are demonstrating that drinking water, food security, and flood control needs can indeed be met in ways that employ rather than destroy ecosystem services. Many are saving their residents a great deal of money by doing so, while creating a healthier environment at the same time. The dike across the Aral Sea stands as a vivid reminder of the urgent need to build on these experiences.

Assessing the Damage— and How We Got Where We Are

It is difficult to imagine today's world of 6.4 billion people and $55 trillion in annual economic output without water engineering—dams to store water, canals to move it from one place to another, pumps to lift water from deep underground, and levees to prevent rivers from flooding valuable property.[1] Hydroelectric dams currently provide 19 percent of the world's electricity.[2] Dams, reservoirs, canals, and millions of groundwater wells have allowed global water use to roughly triple since 1950, bringing supplies to growing cities, industries, and farms.[3] Today, about 40 percent of the world's food comes from the 18 percent of cropland that gets irrigation water.[4] Between 1961 and 2001, engineers and farmers doubled the area of land under irrigation as the "green revolution" package of high-yielding seeds, fertilizers, and water spread to more regions.[5] Rivers that were straightened and deepened for shipping allowed crops and goods to move from continental interiors to ports, expanding trade and prosperity.

These benefits, however, have come at a high price. Human impacts on freshwater systems have reached global proportions and have disrupted a wide range of valuable ecological services.[6] (See Sidebar 2, page 16.) Signs of overstressed and deteriorating ecosystems take many forms—disappearing

FIGURE 2

River Flow Into the Aral Sea, 1926–2003

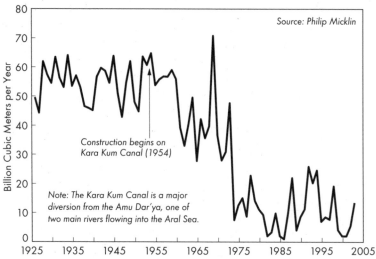

Source: Philip Micklin

Construction begins on Kara Kum Canal (1954)

Note: The Kara Kum Canal is a major diversion from the Amu Dar'ya, one of two main rivers flowing into the Aral Sea.

species, decimated fish populations, falling water tables, altered river flows, shrinking lakes, diminishing wetlands, declining water quality, and pollution-induced "dead zones." Virtually all of these indicators are worsening, and they collectively affect large areas of the globe.

Many major rivers that have been heavily dammed and diverted no longer reach the sea for extended periods of the year. Prior to 1960, the Amu Dar'ya and Syr Dar'ya delivered an average of 55 billion cubic meters of fresh water annually to the Aral Sea, enough to offset net evaporation from the sea's surface.[7] Annual inflows to the sea now average about one-tenth that historical average, causing the sea's steady shrinkage.[8] (See Figure 2.) Along with constructing the dike, Central Asian water managers now hope to restore inflows from the Syr Dar'ya sufficiently to save the Small Aral, which split from the larger sea in 1987.[9]

The Yellow River, China's second largest, ran dry in all but six years between 1972 and 1999.[10] During the 1970s, the average length of dry riverbed was 130 kilometers; by the mid-1990s, it was 700 kilometers. Not only did the length of dry

SIDEBAR 2

Human Impacts on Freshwater Ecosystems and Their Services

Human Activity	Impacts

Land conversion and degradation

Worldwide, half or more of the land in nearly one-third of 106 primary watersheds has been converted to agriculture or urban-industrial uses. Thirteen European watersheds have lost at least 90% of their original vegetative cover. An estimated 25-50% of the world's original wetlands area has been drained for agriculture or other purposes.

- Alters partitioning of rainfall between surface runoff, groundwater recharge, and evapotranspiration.
- Affects quantity, quality, and/or timing of water flows.
- Causes sedimentation of reservoirs.
- Causes habitat degradation and species loss.

Dam construction

Engineers have built more than 45,000 large dams on the world's rivers, up from 5,000 in 1950—an average construction rate of two large dams per day. Dams now affect more than half of the world's large river systems (172 out of 292) and more than three-quarters of large river systems in the United States, Canada, Europe, and the former Soviet Union.

- Fragments rivers and alters natural flow patterns. Dams and reservoirs now intercept about 35% of river flows as they head toward the sea, up from 5% in 1950. They can hold back 15% of annual global runoff all at once.
- Changes water temperature and nutrient and sediment transport. Reservoirs have trapped more than 100 billion tons of sediment that otherwise would have been delivered to coastal regions.
- Blocks fish migration and causes habitat degradation and species loss.

Dike and levee construction

Engineers have diked and channelized thousands of kilometers of rivers worldwide.

- Disconnects rivers from floodplains, eliminating habitat for fish and other aquatic organisms and reducing groundwater recharge.
- Encourages human settlements in floodplains, increasing risk of flood damages.

Large-scale river diversion

River flows have been siphoned off to supply cities and farming regions. A number of large rivers—including the Colorado, Indus, Nile, and Yellow—now discharge little or no water to the sea for extended periods.

- Depletes flows to damaging levels.
- Causes riverine habitat degradation, harm to fisheries, and species loss.
- Reduces water quality.
- Degrades coastal ecosystems and lakes into which rivers empty.

Human Activity	Impacts
Groundwater withdrawals Cities, farmers, and others have over-tapped groundwater in key agricultural regions in Asia, North Africa, the Middle East, and the United States.	• Causes water tables to drop. • Can reduce or eliminate springs and river base flows. • Can deplete underground aquifers.
Uncontrolled pollution Recent decades have seen increased pollution from fertilizer and pesticide runoff, discharges of synthetic chemicals and heavy metals from industry, and releases of acid-forming air pollutants from power plants. Use of nitrogen fertilizer has increased eightfold since 1960.	• Diminishes water quality and safety of drinking water. • Causes habitat and species loss. • Leads to eutrophication and spread of low-oxygen "dead zones." • Alters chemistry of rivers and lakes, destroying habitat, harming fish and wildlife, and increasing risks to human health.
Emissions of climate-altering air pollutants Fossil-fuel burning released more than 7 billion tons of carbon in 2004, nearly three times the amount in 1960. The average atmospheric concentration of carbon dioxide has risen 35% over pre-industrial levels. The 10 warmest years recorded since 1880 have all occurred since 1990.	• Likely to fundamentally alter global water cycle, including shifts in rainfall and river runoff patterns. • Will melt glaciers and shrink snow-packs, reducing future water supplies. • Will alter fish and wildlife habitat. • Likely to increase the number and intensity of floods and droughts.
Introduction of exotic species The spread of non-native species that can colonize ecosystems and alter ecosystem dynamics has increased rapidly with greater movements of people and goods around the world.	• Affects food webs, nutrient cycling, and water quality. • Contributes to species loss. World-wide, at least 20 percent of the world's 10,000 freshwater fish species have become endangered, are threatened with extinction, or have already gone extinct.
Population and consumption growth World population has more than doubled since 1950, to nearly 6.4 billion in 2004. Over this period, global water use has roughly tripled, wood use has more than doubled, and consumption of coal, oil, and natural gas has increased nearly fivefold.	• Places virtually all ecosystem services at greater risk due to increased damming and diverting of rivers, increased land conversion, heightened water and air pollution, and increased potential for climatic change.

Sources: See Endnote 6 for this section.

channel expand over time, but so did the duration of dryness: in 1997, the lower reaches of the Yellow were dry for a record 226 days, causing $1.6 billion in economic damage to Shandong, the province last in line for the river's water.[11] The disappearance of wetlands, harm to aquatic life, and other downstream ecological effects have led the Yellow River Conservancy Commission to restore some minimum flows to the river, and since 2000 the Commission has recorded no periods of zero flow. The overall health of the river, however, remains tenuous.

The Nile and Colorado rivers now deliver little water to their deltas except during years of unusually high floods. Egypt's Mediterranean coastline is retreating because the sediment that used to replenish it is now sequestered behind the Aswan High Dam. That retreat will quicken as the warming global climate raises sea levels. Borg-el-Borellos, a village formerly in the Nile delta, is already two kilometers out to sea.[12] Many deltas around the world are similarly threatened by a combination of sediment-trapping upstream, land subsidence from groundwater and oil extraction, and sea level rise. By 2050, some 28,000 square kilometers of current delta lands could be inundated.[13]

Dams and diversions to power and water the western United States have turned the Colorado River into a meager trickle by the time it reaches its delta in most years. After completion of the Glen Canyon Dam in 1963, the Colorado ran dry in its lower reaches for two decades.[14] (See Figure 3.) The reduction in flows has diminished the Colorado delta's wetlands to a small fraction of their original area, destroyed the fishing and flood-recession farming practiced for centuries by the native Cocopa Indians, brought the desert pupfish and Yuma clapper rail closer to extinction, and reduced shrimp harvests in the upper Sea of Cortez.[15] A massive El Niño flood in the mid-1980s and several smaller floods during the 1990s sent the delta enough water to demonstrate that this ecosystem could come back to life if given water more regularly. But a campaign by citizens and scientists to secure even 1 percent of the Colorado's flow for the delta has so far not met with success.

The details vary, but the story is similar in many other

FIGURE 3

Flow of the Colorado River Below All Major Dams and Diversions, 1904–2004

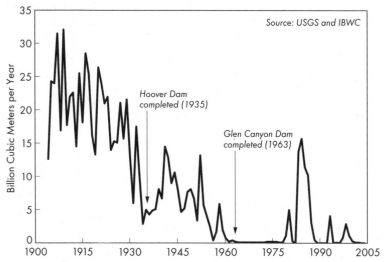

Source: USGS and IBWC

Hoover Dam
completed (1935)

Glen Canyon Dam
completed (1963)

river basins around the world. Dam and diversion projects coming on line in China, India, Turkey, Brazil, and elsewhere—some larger than anything built to date—almost guarantee a host of new ecological wounds. In many of these rivers, the loss of fisheries, biodiversity, and other ecological values will far exceed anything that has occurred before.

Millions of poor people in the developing world still depend directly on river floodplains, delta fisheries, and other natural assets of healthy rivers. As in the Aral Sea region, projects that destroy these ecosystem services can worsen the health and livelihoods of people already living on the economic fringe. In Pakistan, for example, officials have dammed and diverted the Indus River to expand crop production, and 80 percent of the country's cropland is now irrigated.[16] But people downstream in the delta have paid the price. River flows reaching the Indus delta have declined by 90 percent over the last 60 years.[17] Recent droughts have exacerbated the water shortage, leaving the delta and its dependents bereft of fresh water. With so little freshwater discharge to keep the Arabian

Sea at bay, the sea has inundated some 486,000 hectares of delta farmland.[18]

The Indus delta's coastal mangrove forests, which thrive where fresh water mixes with salt water, are also suffering from the lack of river flow. To date, the area of mangroves has shrunk by more than 40 percent, from some 344,000 hectares to 200,000 hectares.[19] Mangroves provide vital spawning habitat for fish and shrimp worth some $20 million per year to coastal dwellers in the region, and they offer storm and wave protection as well.[20] But some of the most prized deltaic fish species have nearly disappeared, and prospects for sustainable livelihoods are dwindling. Hundreds of families have migrated out of the delta region.[21]

In recent years, scientists have determined that altering a river's flow can be just as harmful ecologically as depleting it. Every river has a natural flow regime—a distinct pattern and timing of flow levels determined by the climate, geology, topography, vegetation, and other features of its watershed.[22] In monsoonal climates, for example, river flows peak during the rainy season and then drop to low levels during the dry season. Rivers fed primarily by mountain snowpacks, like the Colorado, will naturally run highest during the spring melting season and lowest during the summer. Periodically—once a decade, or once every half century—extreme floods and droughts may occur that are outside the normal annual flow pattern but that are part of that river's natural regime and critical to its ecological health.[23]

Dams built to intercept, store, and release river runoff when best suited for human purposes have fundamentally altered these natural rhythms. Engineers release reservoir water to hydroelectric turbines when a city needs more power, to agricultural canals when farmers demand irrigation water, and to river channels when barges need to ship goods to and from ports. The flow pattern of the Missouri River, the longest river in the continental United States, for example, now looks nothing like the flow regime that existed before the river was dammed and channelized.[24] (See Figure 4.) Gone are the big early-spring snowmelt floods, the smaller late-spring floods, and

FIGURE 4

Missouri River Flows Before and After Regulation by Dams

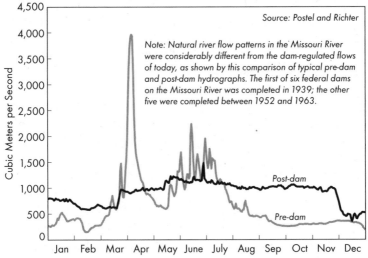

Source: Postel and Richter

Note: Natural river flow patterns in the Missouri River were considerably different from the dam-regulated flows of today, as shown by this comparison of typical pre-dam and post-dam hydrographs. The first of six federal dams on the Missouri River was completed in 1939; the other five were completed between 1952 and 1963.

the summer low-flows, which were sacrificed to maintain sufficient water depth for barge traffic.

These flow modifications have disrupted the habitats to which the myriad varieties of life in the Missouri have become adapted. Fish and other organisms can no longer access floodplains. Sandbars and shallow-water habitats that are critical to fish, birds, and riparian vegetation have disappeared. Flows that provided vital life-cycle cues—a signal to spawn, for example—no longer occur. As a result, numerous species within the Missouri ecosystem are now at risk. Federal or state agencies have listed as endangered, threatened, or rare a total of sixteen species of fish, fourteen birds, seven plants, six insects, four reptiles, three mammals, and two mussels.[25] Production of mayflies, caddis flies, and other invertebrates—key members of the river's food web—has dropped some 70 percent in unchannelized parts of the river.[26] By almost any measure, the Missouri ecosystem is in seriously declining health.

This syndrome of flow modification, habitat destruction, and species imperilment is playing out in river systems all over the world. Although biodiversity loss does not necessarily equate with the loss of ecosystem services, it offers a rough proxy for impacts on ecosystem health and resilience. Overall, the news is not good.

Consider freshwater mussels, which are so effective at purifying water that they are sometimes called "living filters." They suck water in, filter it to get algae and other food out, and then release it—cleaner than before—back to the river, lake, or pond. Unfortunately, freshwater mussels are at great risk because of changes to their habitat from dams, diversions, pollution, and other factors. The United States ranks first in the world in the number of known species of freshwater mussels, and 69 percent of these 292 species are now at risk of extinction or already extinct.[27] The most diverse assemblage of freshwater mussels ever known was located in the middle stretch of the Tennessee River in northern Alabama. Prior to the damming of the river in the early 1900s, 69 mussel species were reported from this reach; since then, 32 of these species have apparently disappeared, with no recorded sightings in nearly a century.[28]

These mussels join other freshwater species whose work on Earth has been brought to a premature end. At least 123 species of North American freshwater fish, mollusks, crayfish, and amphibians have gone extinct since 1900.[29] Biologists Anthony Ricciardi and Joseph Rasmussen estimate that in recent decades, North American freshwater animal species have been extinguished at an average rate of half a percent per decade, and they project this rate to increase in the near future to 3.7 percent per decade—about five times greater than that projected for terrestrial species.

Human impacts on the water cycle are evident not just in rivers, lakes, and groundwater systems, but in coastal zones as well.[30] Rivers nourish coastal waters by transporting nutrients collected from their watersheds to the sea, helping to sustain the highly productive fisheries of coastal bays and estuaries. In recent times, however, many rivers have begun to carry and release an excess of nutrients—principally nitrogen, but also phosphorus—to these coastal areas.[31] The sources of this nutrient enrichment vary, but the main ones include wastewater from heavily populated urban areas, fertilizers from intensively farmed areas, livestock waste from concentrated animal operations, and atmospheric deposition from industrial and automotive air pollutants. In many industrialized regions, riverborne nitrogen has increased up to five-fold from pre-industrial levels.[32] More than half of the coastal bays and estuaries in the United States are now degraded by excessive nutrients.[33]

These nutrient overloads can promote increased algal growth, a process known as eutrophication, which can in turn rob waters of oxygen as bacteria break down excess organic matter. The point at which fish and other animals begin to suffocate varies, but stress is usually evident when oxygen levels drop below 3 milligrams per liter of water. (For comparison, oxygen levels in air are about 280 milligrams per liter.) This state of hypoxia (low oxygen) produces what are often called "dead zones."

Hypoxic areas have spread rapidly in recent decades and now total some 146 worldwide.[34] Because the most productive coastal fisheries are associated with nutrient deliveries from rivers, a good deal of them are at risk from oxygen-depleting over-enrichment. Mass mortalities have occurred in a wide variety of hypoxic areas around the world, including the German Bight of the North Sea, the Kattegat Sea between Sweden and Denmark, the Gulf of Trieste in the Adriatic, the northwest shelf of the Black Sea, most major harbors of Japan, Mobile Bay off the coast of Alabama, and Pamlico Sound off the coast of North Carolina.[35] Vast hypoxic zones exceeding 20,000 square kilometers occur in the Gulf of Mexico, the East China Sea, and the Baltic.[36] In a global review of the problem, Robert Diaz of the Virginia Institute of Marine Science concludes that "no other environmental variable of such ecological importance to estuarine and coastal marine ecosystems around the world has changed so drastically, in such a short period of time, as dissolved oxygen." Oxygen deficiency, he continues, "may very well be the most widespread anthropogenically induced deleterious effect in estuarine and marine environments."[37]

Without serious effort to curb the influx of nutrients to coastal waters, the number and severity of hypoxic areas is bound to increase. Today, more than half of the world's people live within 100 kilometers of a coastline, and this share is expected to rise substantially over the next two decades.[38] Worldwide nitrogen fertilizer consumption has climbed eightfold since 1960 and continues to rise.[39] (See Figure 5.) In Southeast and East Asia, it has risen an astonishing 16-fold since 1960, to 31.8 million tons; today, this region alone accounts for more than one-third of the global total.[40] Consumption in South Asia, now totaling 14.1 million tons, has surpassed that in North America, raising concerns about the spread of dead zones in the coastal waters of Bangladesh, India, and Pakistan.

As if this litany of human impacts on aquatic ecosystems were not enough, there is the wild card of climate change. Scientists warn that climatic changes induced by the buildup of carbon dioxide and other greenhouse gases will fundamentally alter the hydrological cycle.[41] Floods and droughts are expected to intensify as both precipitation and evaporation increase. Floods alone cost the world economy an average of $50–60 billion per year, a sum likely to balloon as floods intensify, more people settle in flood-prone areas, and deforestation fosters more rapid runoff.[42] Crops grown on two-thirds of the world's irrigated land are expected to demand more water because of higher temperatures and evapotranspiration rates, worsening the shortages now looming for irrigated farmland.[43] Although river flows may increase in some areas, they are expected to decrease in many regions already facing shortages, including the Mediterranean, central Asia, the Arabian Peninsula, southwestern North America, parts of Australia, and southern Africa.[44]

Climate change will also disrupt a natural service vital to hundreds of the world's large cities and millions of hectares of irrigated land: water storage. Glaciers and mountain snowpacks are vast natural reservoirs that feed many of the world's great rivers, including those emerging from the Alps, Andes, Cascades, Himalayas, Rockies, and Sierra Nevada. Warmer temperatures will shrink snowfields as more precipitation falls as

FIGURE 5

Nitrogen Fertilizer Consumption, Selected Regions and the World, 1960–2003

rain rather than snow and will cause more rapid melting earlier in the year. The likely upshot is more flooding in the late winter and early spring and lower flows during the late spring and summer, when irrigation demands are greatest. Snowpacks in the Cascade Mountains of Oregon and Washington, for example, may drop as much as 60 percent, reducing summer stream flows by 20–50 percent.[45]

Glaciers are retreating rapidly in many mountainous regions already, including the Alps, the Andes, and the Himalayas. For a period of time, accelerated glacial melting will produce an increase in river runoff—a temporary lift to local water supplies. When the glaciers disappear, however, so will the runoff they generate.

Studies by scientists at Calicut University in Kerala, India, show that the Indus and other great rivers of northern India and Pakistan will run strongly for perhaps 40 years as the glaciers rapidly melt, but then flows will drop off, creating severe water shortages.[46] Flows of the Indus River, for example, could double from the increased melting, but then drop precipitously to half current levels by the end of the century. Likewise, in the Andean country of Bolivia, officials in La Paz now worry about future water shortages because the glaciers that provide the capital city's water are retreating so quickly.[47] Robert Gallaire, a French hydrologist studying the Bolivian glaciers, summed up the problem: "[W]e are using reserves that are being reduced. So we have to ask, what will happen in 50 years? Fifty years, you know, is tomorrow."[48]

Healthy Watersheds for Safe Drinking Water

Bogotá, the capital of Colombia and home to some 7 million people, gets about 70 percent of its drinking water from the *páramo*, a unique high-elevation wetland ecosystem.[1] The vegetation of the páramo absorbs precipitation and snowmelt like a sponge and then releases clean water at a reliable rate of 28 cubic meters per second all year long.[2] The natural filtering action of the wetlands keeps turbidity (a measure of water's cloudiness) and contaminants low. Consequently, there is little need for reservoir storage because the flow is so reliable, nor for expensive treatment because the water is so clean. The raw water delivered from the páramo to the local water plant typically requires only the application of chlorine for disinfection.

This critical watershed lies within the Chingaza National Park, but even so Bogotá's public water utility, Empresa de Acueducto y Alcantarillado de Bogotá (EAAB), works actively to protect it. Throughout much of the Colombian Andes the area of páramo is dwindling rapidly from population and agricultural pressures, threatening the drinking water's reliability and purity.

Even in the face of rapid population growth and civil unrest within the Colombian capital, EAAB managed between 1993 and 2001 to reduce the number of households lacking safe drinking water by 75 percent and to reduce the number without sanitation by more than half, effectively meeting international water and sanitation targets within eight years.[3] Today, 95 percent of Bogotá's households have potable water and 87 percent have sewerage services.[4] Even more impressive, EAAB has dramatically reduced per capita household water use through an effective conservation program, delaying the need to construct new water supply facilities for 20 years.[5] Although serious challenges remain, Bogotá is building water security on the firm foundation of watershed protection, equitable access, and efficient use—a triumvirate that offers the best hope for meeting human needs for safe, affordable drinking water while protecting freshwater ecosystems at the same time.

Worldwide, drinking, cooking, and other residential uses of water account for less than 10 percent of water use. However, because this demand is often highly concentrated in dense urban settings, it can place great pressure on a region's rivers, streams, lakes, and wetlands. Moreover, unlike water for irrigation and many industrial uses, household water must meet high standards of quality. The more polluted the water source, the more expensive it becomes to treat and make safe for drinking. In one extreme case, the city of Mombasa, Kenya, abandoned its water supply system after only a decade of operation because of skyrocketing treatment costs brought on by the deteriorating quality of its source water.[6]

Fortunately, a number of cities and governments around the world are demonstrating the benefits of an underappreciated fact: healthy watersheds are nature's water factories, and it pays to protect them. Forests and wetlands can churn out high-quality water supplies at a lower cost than conventional treatment plants do, while providing many other valuable benefits at the same time, from recreational enjoyment to biodiversity conservation to climate protection.

Several major cities in the United States have avoided the construction of expensive treatment facilities by investing instead in watershed protection to maintain the purity of their drinking water. The U.S. Safe Drinking Water Act requires that cities dependent on rivers, lakes, or other surface waters for their drinking supplies build filtration plants unless they can demonstrate that they are protecting their watersheds sufficiently to satisfy federal water quality standards. Boston, Seattle, and other cities that have taken the watershed protection route are saving their residents hundreds of millions of dollars in avoided capital expenditures.[7] (See Table 1.)

New York City's program is especially interesting because it involves circumstances common to most cities worldwide, including a watershed occupied by a diverse set of landowners and communities rather than one protected and off-limits to use. New York City gets 90 percent of its drinking water from a family of reservoirs in the Catskills-Delaware watershed. More than three-quarters of this watershed is covered by forests, an important asset for water quality protection. However, three-quarters of it is also privately owned, giving the city limited say over how the land is used and managed. Faced with estimated capital costs of $6 billion and annual operation and maintenance costs of at least $300 million for a filtration plant, the city took a bold step: it would attempt to negotiate an agreement with the diverse array of interests in the watershed that included compensation for the important watershed service they were providing to the city—namely, water purification.[8]

After many years of negotiations, New York City signed a landmark agreement in 1997 with state and federal officials, environmental organizations, and some 70 towns and villages committing it to invest on the order of $1.5 billion over 10 years to restore and protect the watershed, along with other measures (including a $60 million trust fund) to improve local economies and livelihoods.[9] The city is acquiring key parcels of land from willing sellers in order to create larger buffers around reservoirs and streams. It has also partnered with organizations in the watershed to reduce polluted runoff from farming practices, promote better forest management, upgrade wastewater infrastructure, and curb streambank erosion and habitat degradation. As of 2004, the city had invested more than $1 billion in the program.[10] In their five-year review, federal authorities found that the program had made sufficient progress to justify extending the filtration waiver. New York City will need to remain vigilant in its efforts,

TABLE 1

Selected U.S. Cities That Have Avoided Construction of Filtration Plants Through Watershed Protection

Metropolitan Area	Population	Avoided Costs Through Watershed Protection
New York City	9 million	$1.5 billion to be spent on watershed protection over 10 years will avoid at least $6 billion in capital costs and $300 million in annual operating costs.
Boston, Massachusetts	2.3 million	$180 million (gross) avoided cost.
Seattle, Washington	1.3 million	$150–200 million (gross) avoided cost.
Portland, Oregon	825,000	$920,000 spent annually to protect watershed is avoiding a $200 million capital cost.
Portland, Maine	160,000	$729,000 spent annually to protect watershed has avoided $25 million in capital costs and $725,000 in operating costs.
Syracuse, New York	150,000	$10 million watershed plan is avoiding $64–76 million in capital costs.
Auburn, Maine	23,000	$570,000 spent to acquire watershed land is avoiding $30 million in capital costs and $750,000 in annual operating costs.

Source: See Endnote 7 for this section.

however, in light of growing population and economic pressures for land development in the watershed.

New York City's program is the largest scheme in effect worldwide in which a beneficiary of a watershed service directly pays the providers of that service. Although the ideal situation may be for a city to own and protect its source watershed, that ideal gets harder to find in today's crowded world. More often, a partnership approach like New York's may be the best bet for sustaining watershed health and protecting valuable drinking water services.

One potentially ripe set of opportunities for increasing investments in watershed protection lies in recognizing the multiple benefits that many watersheds provide. There is substantial overlap, for instance, between lands protected for their biodiversity and conservation values (such as national parks and nature reserves) and lands that supply cities with drinking water. Honduras gave protected status to La Tigra National Park in part because its cloud forests helped generate 40 percent of the water supply for the capital city at a cost equal to some 5 percent of the next best alternative.[11] A 2003 study by the World Bank-WWF Alliance for Forest Conservation and Sustainable Use found that of 105 populous cities in Africa, North and South America, Asia, and Europe, 33 obtain a significant portion of their water supplies from legally protected lands.[12]

Many lands protected on paper, however, may actually be used by local inhabitants for farming, grazing, fuelwood collection, and other activities that can compromise the ability of those lands to continue providing high-quality water. In such cases, a partnership between downstream beneficiaries and upstream land-users, such as one being tried in Quito, Ecuador, may achieve the needed degree of protection while also helping support the livelihoods of people in the watershed.[13] (See Sidebar 3.)

Forests and wetlands are unambiguously good at cleansing water supplies, but their ability to increase the quantity of water available, especially during the critical dry season, varies with local conditions. Popular wisdom has it that planting trees enhances water supplies, but it may in fact have just the opposite effect. Young, fast-growing trees often transpire a great deal of water, especially during the hot, dry season, so planting them in certain landscapes may actually reduce water supplies. In Fiji, for instance, the large-scale planting of pine in watersheds that had previously been grassland resulted in 50-60 percent reductions in dry-season flow—reductions that put both drinking water supplies and the operation of a hydroelectric plant at risk.[14]

South Africa has been working for a decade to reverse the negative water supply and biodiversity impacts caused by the spread of non-native eucalyptus, pine, black wattle, and other thirsty trees into the native *fynbos* (shrubland) watersheds of the Western Cape. The fynbos catchments are part of the Cape Floristic Kingdom, one of only six biogeographic plant

SIDEBAR 3

Protecting Drinking Water Supplies and Biodiversity: The Watershed Trust Fund of Quito, Ecuador

Quito, the capital city of Ecuador and home to more than 1.5 million people, derives about 80 percent of its drinking water from two protected areas, the Cayambe Coca Ecological Reserve and the Antisana Ecological Reserve. These reserves encompass 520,000 hectares of high-altitude grasslands and cloud forests. Although formally protected as part of Ecuador's national park system, the lands are also used for cattle, dairy, and timber production by the 27,000 people living within or around the reserves.

Concern about the impact of these activities on the quantity and quality of water supplied to Quito led to the establishment of a trust fund to finance watershed protection measures. Proposed in 1997 by Fundación Antisana (a Quito-based environmental group), and established in 2000 with support from The Nature Conservancy (a U.S.-based conservation organization) and the U.S. Agency for International Development, the trust fund (Fondo para la Conservacion del Agua, or FONAG) is designed to pool the demand for watershed protection among the various downstream beneficiaries. These include a municipal water supply agency, irrigators, commercial flower plantations, and hydroelectric power stations.

FONAG's non-declining endowment is overseen by an independent financial manager. The fund is designed to receive payments from local water users, as well as donations from international agencies and private organizations. Under the initial contract, Quito's municipal water agency is to contribute 1 percent of its monthly drinking water sales to the fund. Quito's electricity supplier, which generates about 22 percent of its hydropower in the watersheds surrounding the capital, has agreed to pay a flat fee of $45,000 per year. In 2003, a private beer company, Cervecería Andina, committed to contribute $6,000 per year to the fund. As of 2004, FONAG had close to $2 million in capital and was expected to mobilize $585,000 for watershed protection projects—$215,000 from the fund and the remaining $370,000 from other organizations.

An important attribute of FONAG is that it combines biodiversity conservation with water supply protection. This helps both to expand the number of potential contributors and to prioritize areas for project funding. For example, Quito's water agency works with The Nature Conservancy to identify projects in those watersheds that offer both high biodiversity and water supply values. Projects are now underway in such watersheds to improve sheep and cattle production practices in order to reduce impacts on the land and improve water quality.

Sources: See Endnote 13 for this section.

kingdoms in the world and the richest area of endemic plant diversity on the planet. The low-lying fynbos vegetation is well adapted to drought and thrives on relatively little water. With the invasion of the thirsty non-natives, transpiration increased markedly in these watersheds, depleting local streamflows. Not only did the alien invaders threaten the region's amazing plant diversity, they also jeopardized the health of freshwater ecosystems and the sustainability of the water supply.

To nail down these impacts, South African researchers assessed the water supply services provided by healthy native fynbos catchments compared with catchments heavily invaded by alien trees.[15] They found that restoring watersheds to their more native state by removing the alien invaders would yield nearly 30 percent more water than an equivalent-sized watershed without this restoration. Moreover, they estimated that the unit cost of the water supplied from the restored watershed was 14-percent less than that from the degraded watershed— a ringing economic endorsement for alien plant removal.

Through its novel Working for Water Programme, launched in 1995, the South African government has undertaken just such a scheme.[16] The program has trained and employed more than 20,000 people and removed alien plants from more than 1 million hectares of the South African landscape. Not only is the program putting water back into streams and increasing available water supplies, it is creating jobs among the poorest sectors of society, conserving plant species, and helping sustain the flower and tourism industries that depend in part on the nation's unique biodiversity. Despite progress, however, the invasive plants continue to spread at a faster rate than they are cleared, underscoring the importance of preventing aggressive invasives from taking hold.[17]

Meeting drinking water needs in ways that preserve watersheds and the many benefits they provide also requires more concerted efforts to conserve urban water supplies and to use them more efficiently. Cities around the world commonly lose 20–50 percent of their water supplies to leaks in the distribution system and other factors. This water is removed from a natural water source, where it likely performed important ecological work, is treated with chemicals and pumped into the water distribution system, but then leaks away before it reaches a billable customer. Taiwan loses nearly 2 million cubic meters of water a day to leakage, a volume roughly equal to 325 million toilet flushes.[18] Nairobi, the capital of water-stressed Kenya, loses 40 percent of its urban supply to leakage.[19] It makes no sense for a city to build a new dam or an expensive desalination plant when a third or more of its existing supply is leaking away, but this is precisely what many cities do.

By reducing waste and encouraging conservation, cities can leave more water in rivers and lakes, build fewer and smaller dams, pump less groundwater, and reduce the amount of energy and chemicals needed to treat and distribute their supply. Despite the benefits, however, cities still usually view conservation only as an emergency response to drought rather than a core element of water planning.[20] Fortunately, there are shining exceptions to this rule. Copenhagen, Denmark, boasts a system leakage rate of only 3 percent, perhaps the lowest in the world. Fukuoka, Japan, is not far behind, with a 5 percent leakage rate.[21] But no city likely matches the conservation success of the greater Boston area, where water use in 2004 hit a 50-year low.[22] (See Sidebar 4, page 34.)

From Bogotá to Boston, these examples demonstrate that the challenge of providing safe and affordable drinking water can be synchronized with better protection of ecosystem services. In most cases, however, watershed protection remains a neglected stepchild of water supply systems, and conservation is relegated to drought-response at best.

Correcting these oversights is especially important in developing countries, which face simultaneous challenges of reducing rural poverty and meeting the water supply needs of expanding cities and industries, often under very water-stressed conditions. The World Bank could help by incorporating conservation and watershed management into more of its urban water supply projects and by building compensation for natural watershed services into more of its rural development projects. An internal review of the Bank's watershed management

SIDEBAR 4

Conserving Water to Avoid New Dam or Diversion Projects: The Success of Metropolitan Boston

In early 2005, the Massachusetts Water Resources Authority (MWRA), a public agency that serves as water wholesaler for the U.S. city of Boston and more than 40 surrounding cities and towns, revealed a stunning achievement: water demand in the greater metropolitan Boston area had hit a 50-year low, despite substantial economic growth and relatively little change in population. Total water use in 2004 was 295 million cubic meters per year (MCMY), a 31-percent drop from the late-eighties peak of 426 MCMY. (See Figure 6.)

In the mid-1980s, with water demands exceeding the system's safe yield, the water agency began to investigate options to expand the supply. Citizens voiced concern about the environmental impacts of the proposed project, a diversion from the Connecticut River. In response, the agency began an aggressive conservation program in 1987, including the detection and repair of leaks in the system's pipes, the retrofitting of 370,000 homes with efficient plumbing fixtures, industrial audits, meter improvements, and public education. It also helped Massachusetts become the first state in the nation to require low-volume (six-liter) toilets. The upshot was a steep drop in water demand and the indefinite postponement of the river diversion project, saving the MWRA's 2.1 million water supply customers more than $500 million in capital expenditures alone. Holding the lid on demand will require further action, however, as lawn-watering appetites expand in suburban communities.

An important feature of the MWRA system is that the agency funds public participation in water management decisions through a group called the Water Supply Citizens Advisory Committee (WSCAC). Based in Hadley, Massachusetts, in the Connecticut River watershed, WSCAC conducts independent research, holds public meetings regularly, and advises the MWRA on water conservation and watershed protection. Formed in 1977, WSCAC was a major force behind the MWRA's water conservation success and is now focused on watershed protection to help preserve the system's high drinking water quality and avoid the need to build an expensive filtration plant.

Sources: See Endnote 22 for this section.

portfolio from 1990–2000 found, for instance, that 90 percent of the 42 projects examined focused on improving agricultural production and crop yields, but neglected to account for the

Water Use in the Metropolitan Boston Area, 1960–2004

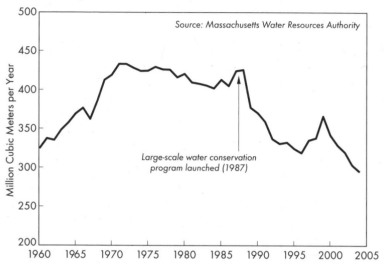

Source: Massachusetts Water Resources Authority

Large-scale water conservation program launched (1987)

additional downstream benefits that better land-use practices and healthier watersheds provide.[23]

Some of the World Bank's own projects show that this narrow focus misses opportunities to link improved drinking water quality downstream with rural poverty reduction upstream. For example, the Bank's micro-watershed project in the Brazilian state of Santa Catarina proved highly successful by most rural development measures: more than 100,000 farmers adopted better land management practices, such as contour terracing and better storage of animal manure; crop yields rose 20–40 percent; farmers outside the project area, inspired by their neighbors, began to upgrade their own practices; and upon completion the project yielded an estimated 20 percent economic rate of return.[24] These measures also produced important water quality benefits, however, that were not captured or secured by the project.

Lauro Bassi, an agricultural engineer based in Santa Catarina, independently studied the water quality effects of this Bank project in the Lajeado São José micro-watershed, which supplies drinking water to the Santa Catarina city of Chapecó.[25] He found a 69-percent reduction in the suspended sediment concentration and a 61-percent reduction in the turbidity of the source water entering Chapecó's treatment plant. This improved water quality allowed for substantial reductions in treatment chemicals, yielding cost-savings of $29,340 per year—savings sufficient to pay back the entire cost of the Lajeado São José micro-watershed project in four years. Indeed, just one year of water treatment cost-savings exceeds the $25,000 paid to project farmers in subsidies to encourage them to adopt land-improvement measures.

Echoing the New York City case, an opportunity exists in this small watershed to strike a deal between the water supplier and upstream farmers that would more equitably split the benefits of the project. Based on its cost-savings, the water supplier should be willing to pay a portion of the funding to upstream farmers in return for the improved water quality and lower water-treatment costs. Such a transaction would transform what is now a short-term project subsidy to participating farmers into a long-term compensation payment for a valuable watershed service: the protection of drinking water quality.

Because most ecosystem services lie outside the realm of commercial markets, governments and public entities have important roles to play in valuing and capturing these benefits. Cities such as Bogotá, Boston, and New York City that have successfully safeguarded the natural water purification services of their watersheds and thereby avoided expensive treatment systems are saving their residents millions of dollars. Governments need to expand these benefits by adopting regulations that require water suppliers, whether public or private, to implement effective watershed protection programs—including measures to protect critical watershed lands and aquifer recharge zones from development, and requirements to build the costs of watershed protection into drinking water prices.

Likewise, government action to reduce waste and improve water-use efficiency is critical. The Ministry of Water in Jordan has set a target of reducing leakage in the domestic water system from the current level of 30 percent to 18 percent by 2015. The move is backed by new studies showing that each *dinar* invested in efficiency improvements would yield 1.9 dinar in health, environmental, and water supply benefits.[26] In the United States, federal efficiency standards for toilets, urinals, faucets, and showerheads came into effect in 1997; by 2020, these efficiency improvements are projected to save 8.4–12.4 billion cubic meters of water per year, enough water to supply four to six cities the size of New York City.[27]

Protecting the reliability and safety of drinking water becomes ever more challenging as new sources and types of pollutants enter watersheds. Many chemicals used in everyday products—including sunscreens, plastics, and cosmetics—as well as lawn pesticides and prescription drugs are turning up in water supplies. In August 2004, scientists found the anti-depressant drug Prozac in rivers and groundwater used for drinking supplies in Great Britain.[28] In the United States, the U.S. Geological Survey has found that 80 percent of 139 streams sampled in 30 states contain traces of at least one drug, endocrine-disrupting hormone, insecticide, or other chemical—some at levels that have been shown to harm fish and other aquatic life.[29]

Researchers at Texas Tech University reported in 2005 that they found perchlorate, a chemical used in explosives and rocket fuel, in 99 percent of milk samples taken from a number of U.S. states.[30] The average concentration in breast milk (10.5 micrograms per liter) was 5 times higher than in dairy milk and 10 times higher than the maximum level the U.S. Environmental Protection Agency had sought for drinking water. Perchlorate inhibits uptake of iodine and reduces production of thyroid hormone. High concentrations in breast milk increase the risk of harmful developmental effects in breast-feeding infants. Perchlorate is present in drinking water supplied to at least 11 million people and enters the nation's food supply through contaminated irrigation water.[31] Known releases of the chemical have occurred widely throughout the country, many from military facilities.[32] Virtually the entire lower Colorado River, which irrigates a significant share of the nation's lettuce and other vegetables, is contaminated with perchlorate.

Neither healthy watersheds nor modern treatment facilities can remove all potentially hazardous substances from water, which places a premium on preventing them from entering the environment in the first place. The myriad signs of health effects on wildlife and people from exposure to pesticides and other synthetic chemicals, publicized widely in 1962 with Rachel Carson's *Silent Spring* and again in 1996 in the ground-breaking book *Our Stolen Future*, have raised alarms but have not stopped the flood of chemicals into the environment.[33] Worldwide, 100,000 synthetic chemicals are on the market and 1,000 new ones are introduced each year, most of them without adequate testing and review for their toxic, cancer-causing, endocrine-disrupting, and reproductive effects in people and wildlife.[34]

Food Security With Ecosystem Security

On an average day, a typical resident of the United States consumes more than 5,000 liters of water, not by drinking this quantity out of a glass, but by eating the typical U.S. diet.[1] Wheat, rice, corn, and other crops naturally consume a great deal of water during their growth. Producing a kilogram of grain requires 1,000–3,000 liters of water, whether from rainfall, irrigation, or both.[2] When, as in the United States, diets include a large share of animal products—especially grain-fed beef, which can require 20 times more water per calorie than wheat does—dietary water consumption climbs upward.[3]

These numbers give a glimpse of the challenge of satisfying the food demands of the growing human population while at the same time sustaining freshwater and terrestrial ecosystems. Already, as much as 10 percent of global food production depends on the overpumping of groundwater.[4] In India, where millions of wells have run dry, that figure is closer to 25 percent.[5] These hydrological deficits create a bubble in the food economy that is bound to burst, and they raise questions about where the additional water needed for future food production will come from.

Any hope of satisfying the dietary needs of 8 billion people without dangerous harm to freshwater ecosystems will require roughly a doubling of agricultural water productivity, which means getting twice as much dietary benefit out of every liter of water extracted or appropriated from natural ecosystems for crop production.[6] The most promising opportunities for achieving this doubling lie in three areas. The first is storing, delivering, and applying irrigation water more efficiently. The second is increasing harvests from lands watered only by rainfall. And the third is reducing the amount of meat in diets to healthier and more sustainable levels. These measures are also key to reducing chronic hunger, which now saps the health and energy of 852 million people and kills more than 5 million children each year.[7]

Currently, about 40 percent of the world's food comes from the 18 percent of global cropland that gets irrigation water.[8] For five thousand years, irrigation has been a cornerstone of food security, and it remains so today. However, agriculture accounts for the lion's share of the water withdrawn from rivers and aquifers—nearly 70 percent globally and as much as 90 percent in many developing countries—setting up intense competition for water between irrigated farms, expanding cities, and stressed ecosystems.[9]

Fortunately, there is great potential to improve irrigation efficiency. About half of the water captured by dams, stored in reservoirs, delivered through canals, and applied to fields never actually benefits a crop.[10] Some of this water is lost to evaporation (12 percent of the Nile River evaporates from Egypt's Lake Nasser, for example), some of it seeps through canals and recharges groundwater, and some runs off a farmer's field to a nearby stream, where another farmer downstream can use it.[11] But regardless of which pathway the water takes, these inefficiencies cause more and larger dams and diversions to be built, more salinization of farmland, more pollution of rivers and groundwater with agricultural chemicals, and generally more harm to aquatic ecosystems.

By improving irrigation efficiency, farmers and engineers can free up a substantial amount of water to sustain freshwater ecosystems. On a global basis, curbing current irrigation withdrawals by 10 percent would save enough water to meet the estimated additional water demands of cities and industries by 2025.[12] A varied menu of options exists for achieving such savings, ranging from lining delivery canals, scheduling irrigations to better match crop water needs, applying water more directly to crop root zones, and capturing and reusing water that runs off the field, to name just a few.[13]

Drip and other micro-irrigation methods, for example, deliver precise amounts of water more directly to the roots of plants. They can reduce the volume of water applied to fields by 30–70 percent, while increasing crop yields by 20–90 percent—resulting in a doubling or tripling of water productivity over conventional methods.[14] Worldwide, micro-irrigation is used on about 3.2 million hectares, just over 1 percent of irrigated land. Pervasive and heavy subsidies that have kept irrigation water prices low have constrained the adoption of these efficient methods and encouraged waste instead. But in countries such as Israel and Jordan, micro-irrigation is used extensively, and in others it is expanding at a rapid clip: over the last decade, the area under micro-irrigation has more than doubled in Mexico and South Africa and more than tripled in Spain.

The emergence of low-cost drip systems designed for poor farmers holds great promise not only for increasing yields in water-short areas, but for lifting poor rural families out of poverty. Simple drip systems costing less than one-tenth as much as the conventional U.S. varieties have begun to catch on in India and elsewhere as farmers seek higher yields from limited water supplies.[15] Paul Polak, president of Colorado-based International Development Enterprises, which has developed a low-cost system called Easy Drip, estimates that 200,000 farmers are now using low-cost drip systems worldwide.[16] In

India, farmers are turning to them in large numbers in water-stressed states such as Gujarat and Maharashtra. According to Polak, annual sales of low-cost drip systems in India have climbed to $4 million in just three years. With perhaps 20–30 percent of India's drip-irrigated land now in low-cost systems, drip irrigation is helping raise not only water productivity but incomes for India's poor farmers.

Harmonizing food production with the protection of freshwater ecosystems requires special attention to rice, the preferred staple of about half the human population. More than 90 percent of the world's rice is produced in Asia, where many rivers and aquifers are already overtapped. Rice is typically grown in a layer of standing water 5–10 centimeters deep, which not only requires a great deal of water, but can render soils less fertile for wheat and other crops grown in sequence with rice, as well as add more methane, a potent greenhouse gas, to the atmosphere. According to Prem Bindraban and Huib Hengsdijk of the Netherlands-based Plant Research International, producing a kilogram of irrigated rice can take 2,000-5,000 liters of water, compared with a theoretical minimum requirement of 600 liters.[17]

Many studies have shown, however, that keeping rice fields flooded throughout the growing season is not essential for high yields.[18] Farmers can apply a thinner layer of water or allow rice fields to dry out between irrigations, thereby reducing water applications by 10–70 percent (depending upon local conditions) without a significant drop in yield. One study in the Citarum basin of Indonesia found that such a shift to water-conserving rice cultivation could increase water supplies for nearby urban areas by 40 percent, lessening both urban-rural competition for water and pressures on freshwater ecosystems.[19] Researchers at Cornell University and the Association Tefy Saina are advancing a System of Rice Intensification (SRI) that increases the productivity of irrigated rice through better management of plants, soils, water, and nutrients.[20] SRI has produced water savings of 50 percent and yield increases of 50–100 percent. It has quadrupled rice yields in parts of Madagascar, where soils are among the poorest in the world, from an average of 2 tons per hectare to 8 tons per hectare, while saving water.

The second piece of the water-for-diets challenge—raising water productivity on rainfed lands—is critical not only to lifting global food production but to alleviating hunger and poverty. The vast majority of the world's 852 million hungry people live in rural areas, many of them on small farms in sub-Saharan Africa and South Asia.[21] Some 500 million live in dry or semi-arid areas, where lack of reliable rains severely limits crop production. Besides increasing food security and incomes, higher yields on rainfed lands reduces the total area of land needed for food production, sparing more forests from the axe and chain saw.

Rainfed croplands are often neglected by water planners and managers because they do not involve the operation of dams, canals, wells, and pumps; they rely instead on natural rainfall.[22] But just as irrigation water follows different pathways, so too does rainwater. It may pool on leaf and soil surfaces and evaporate back to the atmosphere without benefiting a plant; or it may replenish the root zones of soils, where plants take it up and transpire it back to the atmosphere. It may also seep through the soil to recharge groundwater, or run off the land to join the flow of a nearby stream. The challenge is to get more benefit per drop of rainfall.

Much of the rain falling on croplands already does important work: about 60 percent of the world's food comes from rainfed lands.[23] But in many areas, 30–50 percent of rainwater evaporates back to the atmosphere without aiding plant growth. (Evaporation itself serves a useful purpose, however, because it recycles moisture back to the atmosphere and helps sustain rainfall.) In southern Africa, for example, it is common for 65 percent of rainwater to evaporate quickly back to the atmosphere, 15 percent to recharge rivers and aquifers, and only 20 percent to get used for transpiration by crops.[24] Similarly, studies of barley fields in northern Syria have found that transpiration accounts for less than 35 percent of rainfall, while those in semi-arid Gansu Province in China have found that only 15–20 percent of rainfall becomes productive transpiration.[25]

Harvests in such regions are generally low because crops do not get enough water. In the semi-arid savannas, which lie between the forests and deserts of tropical regions (including, for instance, the African Sahel), it is not uncommon for croplands to yield as little as 0.5–1.0 ton per hectare, less than a small family needs to lead active, healthy lives. These savannas span roughly 40 percent of the world's land area, and a large proportion of the world's poor and hungry live in them.[26]

Just as a menu of options exists to raise irrigation water productivity, one also exists for improving rainwater productivity.[27] The best choices will vary with local soil, climate, cultural, and other conditions, so farm communities will often need information and technical assistance to tailor measures to their situations. These measures can include sowing seeds early, intercropping to create more canopy cover, selecting deep-rooted crop varieties, cultivating soils to promote more infiltration of rainwater, mulching fields to retain more moisture, and controlling weeds—all of which can increase the amount of crop per drop of rainfall. In addition, farm communities can capture a portion of the local rainfall that would otherwise run off the land and store it to irrigate their crops during dry spells. These "water harvesting" techniques not only boost yields, but can prevent total crop failure. For poor farmers, who are by necessity risk-averse, the possibility that a dry spell will destroy an entire harvest often prevents them from investing in better crop varieties, fertilizers, and other yield-enhancing inputs.

Water harvesting is an ancient practice that can take many forms, but it essentially involves channeling rainwater into ponds, shallow aquifers, or other storage locations for later irrigation use. Researchers have found that combining supplemental irrigation with judicious application of fertilizer can nearly triple sorghum yields in the poor African Sahelian country of Burkina Faso, from 0.5 to 1.4 tons per hectare, and more than double water productivity.[28] Similar studies in a slightly more favorable soil and climatic setting in Kenya found that average maize yields increased 70 percent, from 1.3 tons per hectare to 2.2 tons, and water productivity rose 43 percent.[29] When combined with low-cost drip or other water-thrifty irrigation practices, water harvesting can boost water productivity even more.

The sustainable cultivation of wetlands offers another opportunity to improve food security while protecting freshwater ecosystem services. Long viewed as "wastelands" to be drained and converted to intensive farming or urban development, wetlands are now known to provide more eco-service value per hectare than most other ecosystem types. They store and purify water, mitigate floods and droughts, provide critical spawning and feeding habitat for fish and wildlife, and harbor a rich variety of species. By "reclaiming" wetlands, societies might gain additional land for agriculture or human settlement, but they lose valuable benefits and services that nature provides for free.

In many parts of the developing world, wetlands also enhance food security by supporting a "hungry season" harvest. Farmers draw on the moisture stored in wetland soils to grow crops when there is insufficient rain to support dryland production. Researchers are finding that with careful soil and water management, wetland cultivation can provide this extra food security without overly sacrificing valuable ecological services.[30] In the western highlands of Ethiopia, where in some locales wetland agriculture appears to have been practiced for more than 250 years, communities have followed indigenously-derived principles for sustaining multiple uses of wetlands.[31] These include leaving portions of the wetland undrained, avoiding the overdraining of any areas, protecting springs at the head and outlet of the wetland, maintaining the annual flood regime, limiting or banning cattle grazing, and maintaining a well-vegetated catchment.

In the Pallisa District of Uganda, wetlands spanning 71,100 hectares are used by the local people for rice production and animal grazing as well as for fish, wild vegetables, medicines, and building materials. The wetlands also provide flood control, water purification, and year-round water supplies for the surrounding towns and villages. All together these wetlands are estimated to provide goods and services worth more than

$34 million per year, or nearly $500 per hectare.[32] Because these ecosystem services are not taken into account in conventional decision making, however, Pallisa's wetlands continue to be viewed as wastelands and "reclaimed" for other uses.

To be sure, managing wetlands to grow crops while also sustaining their many other benefits is not easy; the potential for degradation and loss of natural wetland functions is high. But the payoffs from successful efforts are high, too: greater food security for local farm families and sustained ecosystem services for a wider population. Moreover, when added up, the benefits of these varied uses of wetlands can greatly exceed those provided by conventional dam and irrigation projects that, once built, often end up destroying those wetlands. A case in point comes from northeastern Nigeria.[33] (See Sidebar 5, page 46.)

Dietary choices have an important role to play in the task of doubling water productivity as well. Foods vary greatly both in the amount of water they take to produce and in the amount of nutrition—including energy, protein, vitamins, and iron—they provide. For example, it can take 5 times more water to supply 10 grams of protein from beef than from rice, and nearly 20 times more water to supply 500 calories from beef than from rice.[34] (See Figure 7, page 47.) These disparities create opportunities to meet food needs in more ecologically sustainable ways by adjusting diets. While the nearly 1 billion people in the world who are undernourished need to consume more food to lead healthy, productive lives, those at the high end of the diet spectrum can improve their own health as well as the planet's by shifting diets partially away from water-intensive animal products.

The average North American diet, for example, takes nearly twice as much water to produce as an equally (or more) nutritious diet low in meat consumption. An average U.S. resident who decides to reduce the intake of animal products by half would lower the water intensity of his or her diet by 37 percent.[35] If all U.S. residents made such a shift by 2025, when the U.S. population is projected to reach 350 million, the nation's total dietary water requirement would drop by 256 billion cubic meters per year, a savings equal to the annual flow

SIDEBAR 5

Valuing Water for Food, Livelihood, and Ecosystem Security: The Hadejia and Jama'are Floodplains of Northeastern Nigeria

An extensive floodplain spans part of northeastern Nigeria at the confluence of the Hadejia and Jama'are rivers in the Lake Chad watershed. Many rural Nigerians depend on this floodplain for food and income. They use it to graze animals, grow crops, collect fuelwood, and catch fish. The floodplain recharges regional aquifers, which are vital water supplies in times of drought. The Hadejia-Jama'are wetlands also provide dry-season grazing for semi-nomadic pastoralists and critical habitat for migratory waterfowl.

Like many floodplain wetlands, those of the Hadejia and Jama'are rivers are increasingly threatened by existing and proposed dams and irrigation schemes. To help gauge the impact of such projects, researchers Edward Barbier and Julian Thompson evaluated the economic benefits of direct uses of the floodplain—specifically for agriculture, fuelwood, and fishing—and compared these with the economic benefits of the irrigation projects. They found that the present value of the net economic benefits provided by use of the natural floodplain exceeded those of the irrigation project by more than 60-fold (analyzed over both 30 and 50 years). They also determined that water had an economic value in its floodplain uses totaling $9,600 to $14,500 per cubic meter, compared with only $26 to $40 per cubic meter for the irrigation project. Had Barbier and Thompson been able to estimate the value of natural habitat, groundwater recharge, and other ecosystem benefits provided by the intact floodplain, the disparity between the options would have been even greater.

Source: See Endnote 33 for this section.

---

of 14 Colorado Rivers and enough water to meet the dietary needs of more than 200 million people.[36]

Together, getting more nutritional benefit from the water devoted to food production, raising the productivity of both irrigation water and rainfall, and valuing the multiple benefits of water in wetlands and other ecosystems could go a long way toward meeting the world's dietary needs while minimizing harm to already-stressed ecosystems. These adjustments will not happen, however, without policies to encour-

Water Consumed to Supply Protein and Calories

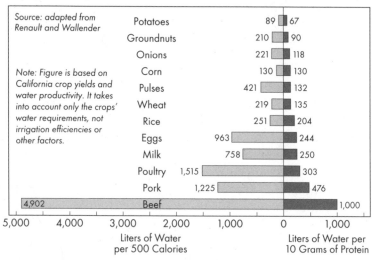

Source: adapted from Renault and Wallender

Note: Figure is based on California crop yields and water productivity. It takes into account only the crops' water requirements, not irrigation efficiencies or other factors.

Crop	Liters of Water per 500 Calories	Liters of Water per 10 Grams of Protein
Potatoes	89	67
Groundnuts	210	90
Onions	221	118
Corn	130	130
Pulses	421	132
Wheat	219	135
Rice	251	204
Eggs	963	244
Milk	758	250
Poultry	1,515	303
Pork	1,225	476
Beef	4,902	1,000

age them. Research, development, and training focused on raising the productivity of rainfed agricultural lands in poor countries is especially important. The food security of 1 billion people and the conservation of remaining terrestrial and freshwater ecosystems depend on this.

Reducing Risks, Preserving Resilience

Nearly 5,000 Haitians lost their lives and tens of thousands lost their homes during tropical storms in May and September of 2004.[1] Although tagged as natural disasters, these tragedies were exacerbated by a distinctly human activity: the clearing of trees in the Haitian highlands. Destitute and lacking alternatives, Haiti's poor have cut down most of their trees for fuelwood and charcoal. In doing so, they have lost a valuable service provided by forested watersheds—the moderation of local flood runoff and the prevention of massive mudslides.[2] Indeed, the same storms that devastated Haiti had far less impact on neighboring Puerto Rico, where the highland watersheds are mostly forested.[3]

Several months later, on December 26, 2004, the tsunami that struck coastal Asian nations and claimed some 273,000 lives cast a spotlight on another valuable ecosystem service—the storm and wave protection afforded by mangroves and coral reefs.[4] The tangled roots and dense vegetation of mangroves, which thrive where salt water meets fresh water, act like a shock absorber against storm and wave energy. Vast areas of these natural protective barriers had been cleared for hotels, shrimp farms, and other commercial developments, including half the coastal mangroves in Thailand. It is impossible to know how many lives might have been spared had these natural buffers remained in place. Some observers believe that tens of thousands of lives were indeed saved in India, Malaysia, and Sri Lanka because of the conservation of mangrove forests and coral reefs there.[5] Within a month of the tragic disaster, officials in Indonesia—where more than 126,000 of the tsunami deaths had occurred and where 30 percent of the coastal mangroves (some 650,000 hectares) had been lost in the preceding few decades—announced a large-scale effort to restore the nation's mangrove defenses.[6]

Although the Asian tsunami sent the annual toll of disaster deaths soaring in 2004, the loss of life and property due to natural disasters has been climbing for two decades. According to data collected by Munich Re, one of the largest insurers of insurance companies, economic losses from natural catastrophes during the last 10 years have totaled $566.8 billion, exceeding the combined losses from 1950 through 1989.[7] More than four times as many "great" natural catastrophes occurred during the 1990s as during the 1950s.

Distinguishing a natural disaster from a human-induced one is getting more difficult. Storms, floods, earthquakes, and tidal waves are natural events, but the degree to which they produce disastrous outcomes is now often strongly influenced by human actions. By necessity or choice, more people are living along coastlines, in floodplains, and on fragile hillsides—

zones that place them in harm's way. At the same time, the clearing of trees, filling of wetlands, engineering of rivers, and destruction of mangroves has frayed the natural safety nets that healthy ecosystems provide. Consequently, when a natural disaster strikes, the risks of catastrophic losses are higher. And global warming from the buildup of greenhouse gases practically guarantees that the number and intensity of damaging storms will increase in the years ahead, a trend that may already be under way.[8]

The high risk to life and property from this confluence of disaster-producing circumstances places a premium on preserving what remains of nature's protective infrastructure and restoring more of it where possible. Just as Indonesia is looking to reestablish its coastal mangrove forests, so El Salvador, Guatemala, and Venezuela have initiated watershed protection programs following devastating storm damages in the late 1990s.[9] Hurricane Mitch, which battered Central America in October 1998 and dumped two meters of rain on the region, claimed some 10,000 lives and caused economic damages conservatively estimated at $8.5 billion, a significant share of the combined GDPs of the worst-affected countries.[10] Although hurricanes are common in this region, the high rates of deforestation in mountainous watersheds almost certainly exacerbated local flooding and mudslides, which swept away homes, roads, bridges, and people.

Large floods have also caused serious damage in parts of Europe and the United States over the last 15 years. Dams, levees, and the straightening of channels have disconnected many rivers in these regions from their floodplains, replacing natural flood protections with engineered ones. The draining of wetlands and the establishment of farms and towns in floodplain zones typically follow these engineering projects, setting the stage for costly disasters. In periods of intense rainfall, these artificial protections can fail to protect as raging rivers bypass levees and attempt to win back their floodplains. Between 1998 and 2002, European countries experienced 100 major floods—including extreme events along the Danube and Elbe rivers—that collectively caused 700 deaths, the displacement of half a million people, and economic losses totaling 25 billion Euros.[11]

More than 10 million people live in areas at risk of extreme floods along Europe's Rhine River, a heavily channelized waterway that no longer meanders but flows artificially straight between engineered embankments.[12] In its upper reaches, the river is cut off from 90 percent of its original floodplain.[13] The Rhine now flows twice as fast as it did before, and flooding in the basin has grown more frequent and damaging. The European Commission estimates that assets worth 165 billion Euros are potentially at risk from flooding of the Rhine.[14]

In light of these damages and risks, some governments have begun to re-evaluate their approaches to flood protection. Along the Danube River, whose watershed includes 14 countries and some 80 million people, collaborative efforts are under way to restore dried-out floodplains and delta wetlands.[15] The governments of Bulgaria, Romania, Moldova, and Ukraine have pledged to create a network of at least 600,000 hectares of floodplain habitat along the lower Danube, the Prut River, and in the Danube Delta. This joint effort is part of a project called Green Corridor for the Danube initiated in June 2000 by the World Wide Fund for Nature (WWF), a private conservation organization.[16] With funding from United Nations agencies and others, the project aims initially to demonstrate how healthy floodplains can not only mitigate flood damages but also provide habitat, reduce pollution loads, and enhance fisheries. According to one estimate, a $275 million investment in wetland restoration in Romania alone would be recouped within six years from the ecosystem goods and services provided by the revitalized delta.[17]

In the United States, the Great Midwest Flood of 1993 provided motivation to re-think flood management. Following intense rains that year, the upper Mississippi and Missouri rivers rose to record heights, and their floodwaters breached levees spanning some 10,000 kilometers.[18] After this episode, researchers estimated that restoration of 5.3 million hectares of wetlands in the upper Mississippi River basin, at a cost of some $2–3 billion, would have absorbed enough floodwater to substantially reduce the flood damage, which was valued at $16–19 billion.[19]

Projects undertaken by the U.S. Army Corps of Engineers, the nation's premier builder of flood control dams and levees, have also shown that wetlands and natural floodplains can do the job at a lower cost than engineering approaches, while providing numerous side benefits. For example, several decades ago the Corps purchased 3,440 hectares of floodplain wetlands in the upper reaches of the Charles River watershed in eastern Massachusetts. The Corps had calculated that with these additional wetlands the floodplain could store 62 million cubic meters of water—roughly equivalent to the storage capacity of a proposed dam. Purchasing the development rights to the wetlands cost $10 million, just one-tenth the $100 million estimated cost of the proposed dam-and-levee project.[20]

More recently, the Corps signed on to a citizen-led proposal to reconnect California's Napa River with its floodplain. The project involves relocating homes, roads, and businesses situated in the natural floodplain zone and revitalizing seasonal wetlands and marshlands. The project's estimated cost of $155 million is one-tenth that of the estimated $1.6 billion that would be needed to repair flood damage over the next century if the floodplain is not restored.[21] Napa County residents voted in 1998 to raise their local sales tax to help pay for the project.[22] They saw this cost as a good bargain for the full range of benefits that would result, which included not only flood control, but reduced flood insurance rates, parks and trails for recreation and bird-watching, higher tourism revenues, and a revitalized downtown. Delegations from Argentina, Australia, and China have already visited the Napa restoration effort to learn from the experience.

Healthy watersheds, floodplains, wetlands, and mangroves are critical pieces of nature's infrastructure. In addition to producing tangible goods and services, they provide disaster insurance—protection against catastrophic losses when natural disasters strike. People buy home insurance to guard against the potential loss of property due to fires and floods, and they buy life insurance to protect their families from a catastrophic death. Such investments cost money now but safeguard families from potential large losses in the future. In a similar way, societies that invest now to protect watersheds, floodplains, mangroves, and other risk-reducing natural infrastructure can avert catastrophic losses later. A risk-taking gambler might say that technology will substitute adequately for ecosystem services over the long term, so we need not bother to conserve them now. If the gambler is wrong, however, the consequence is catastrophic loss—precisely the outcome society wishes to avoid.

Nature's way of mitigating disasters also has a robustness and resilience to it that technology alone cannot offer. Global warming and its anticipated effects on the hydrological cycle will make these attributes all the more important, as tropical storms, spring flooding, and seasonal droughts increase in frequency and intensity. The cautious approach—to avoid catastrophic outcomes by acquiring disaster insurance through ecosystem protection—also sustains fisheries, safeguards biodiversity, and purifies drinking water supplies at the same time.

Bringing Water Policies Into the 21st Century

Few realms of policy making are more out of sync with modern realities than that of fresh water. Signs of water scarcity and ecosystem disruption are pervasive and spreading, yet policies continue to promote inefficient, unproductive, and ecologically harmful practices. Heavy subsidies for irrigation water encourage waste rather than efficiency. Unregulated pumping of groundwater drives water tables ever lower and aquifers closer to depletion. Large dams and diversions intercept more river flows and dry up more wetlands, harming downstream populations and ecosystems while often failing to provide their promised benefits. It almost seems as if the point of public policy is to liquidate Earth's water assets like a store going out of business.

The water policies of the twentieth century helped supply drinking water, food, electricity, and flood control to a large proportion of the human population. But they failed to distribute those benefits equitably, and they largely ignored the role of freshwater ecosystems in sustaining goods and services of great value to society. Just because the marketplace does not assign a price to something does not mean that it lacks worth. Indeed, as examples in earlier sections of this paper have shown, the total value of intact wetlands, floodplains, deltas, and watersheds can exceed by several times the economic value of the activity for which they are sacrificed. In addition, employing nature's services can help purify drinking water, alleviate hunger, mitigate flood damages, and meet other societal goals at a fraction of the cost—sometimes one-tenth to one-half the cost—of conventional technological alternatives.

These realities call out for an overhaul of water policy and a new framework for decision making. The old dam-and-divert mindset viewed freshwater ecosystems as resources that acquired value only when they were engineered and exploited. A mindset in sync with modern knowledge, however, incorporates the value of ecosystems in their unmodified state and aims to employ nature's services for society's benefit. Because every decision to alter an ecosystem results in the loss of goods and services—water quality, fisheries, flood control, species diversity—this new mindset is essential to making informed choices and weighing tradeoffs.

At the heart of this shift in policy must be a reaffirmation of the public trust—the recognition that governments hold certain rights and entitlements in trust for the people and are obliged to protect them for the common good. With the winds of privatization and globalization knocking harder at every door, governments need to assert more forcefully that they hold entitlements to water that have priority over commercial enterprises. These entitlements provide benefits to society that conventional markets do not price and therefore will not protect. Governments that sell these entitlements to the highest bidder violate the public trust.

Leadership on how to translate a public-trust philosophy of water management into policy and practice is emerging from an unlikely place: South Africa. After it came to power in 1994, Nelson Mandela's post-apartheid government undertook a re-writing of the country's constitution and laws, which included passage of a new National Water Act in 1998.[1] Grounded firmly in the doctrine of the public trust, this law establishes a Water Reserve consisting of two parts. The first is a non-negotiable water allocation to meet the basic drinking, cooking, and sanitary needs of all South Africans. (When the government changed hands, some 14 million poor South Africans lacked water for these basic needs.) The second part of the Reserve is an allocation of water to support ecosystem functions so as to secure the valuable services they provide to South Africans. Specifically, the Act says, "the quantity, quality and reliability of water required to maintain the ecological functions on which humans depend shall be reserved so that the human use of water does not individually or cumulatively compromise the long term sustainability of aquatic and associated ecosystems."[2] The water determined to constitute this two-part Reserve has priority over licensed uses, such as irrigation, and only this water is guaranteed as a right.

Following South Africa's lead, a number of national and international conferences, commissions, legislative directives, and laws have called for similar approaches. Importantly, at the December 2001 International Conference on Freshwater in Bonn, Germany, delegates from 118 countries included in their recommendations to the following year's World Summit on Sustainable Development that "the value of ecosystems should be recognized (sic) in water allocation and river basin management," and that "allocations should at a minimum ensure flows through ecosystems at levels that maintain their integrity."[3] The Millennium Ecosystem Assessment, the four-year effort completed in 2005 and carried out under the auspices of the United Nations by some 1,360 scientists, also places the determination of ecosystem water requirements as a high priority.[4] In effect, these are clarion calls to the world's governments to overhaul water policies so as to safeguard freshwater ecosystems.

The first critical step is for governments to establish boundaries or caps on the degree to which human actions harm watersheds, river systems, and groundwater.[5] For rivers already heavily regulated by dams, for example, this means developing a schedule of reservoir releases that mimics the river's natural flow regime while still accommodating economic uses of the river. For overtapped rivers, it means capping withdrawals and returning some water to the river. These caps are not anti-development, but rather pro *sustainable* development: when based on good scientific knowledge, they help ensure that vital ecosystem functions are sustained in the midst of economic growth. They also unleash the power of conservation, efficiency, and markets to raise the productivity of water.

Caps come in many forms and go by many names, but they are in most cases the critical missing piece in water management today.[6] (See Table 2, page 56.) In Australia, for example, water withdrawals from the Murray-Darling river basin, the nation's largest and most economically important, have been capped in order to arrest the severe deterioration of that river system's health.[7] After a tripling of withdrawals between 1944 and 1994, river flows dropped to ecologically harmful levels. Wetlands and fish populations decreased, while salinity levels and the frequency of algal blooms increased. Severe low-flows now occur in the lower Murray River in 60 percent of years, compared with 5 percent under natural conditions. The Murray's flow dropped so low in 2003, a drought year, that its mouth became clogged with sand.[8]

The Murray-Darling watershed spans parts of four states (New South Wales, Queensland, South Australia, and Victoria) and all of the Australian Capital Territory. Through the Murray-Darling Basin Commission (MDBC), these political entities work cooperatively to manage the river. In 1997, in response

TABLE 2

Caps on the Modification of Freshwater Ecosystems, Selected Examples

Ecosystem/Region	Nature of the Cap
Murray-Darling River Basin, Australia	Multi-state river basin commission established a cap on water withdrawals in 1997 to arrest degradation of river system.
Great Lakes, United States and Canada	2001 Annex to the Great Lakes Charter calls for no net degradation of the basin's freshwater ecosystems; rules for implementation still under discussion.
European waters, European Union	2000 Water Framework Directive establishes criteria for classifying the ecological status of water bodies; calls on member countries to prevent any deterioration in this status and to bring all to at least "good."
Ipswich River, Massachusetts, United States	State officials set mandatory withdrawal restrictions on each town permitted to use the river; when flows drop to a specified level, communities must institute water conservation measures.
Yellow River, China	River commission must reduce water diversions when flow drops to a specified level to prevent the river from running dry.
Edwards Aquifer, Texas, United States	State legislature capped pumping from the aquifer to sustain flows to surface springs that support endangered species.
Pamlico Estuary, North Carolina, United States	State officials set targets for discharges of phosphorus and nitrogen into the estuary and allowed for trading of nutrient credits to meet goals cost-effectively.

Sources: See Endnote 6 for this section.

to the rapid deterioration of the river's health, the Minister-ial Council (which consists of resource ministers from each basin state or territory plus the Commonwealth) placed a cap on diversions from the basin. According to the MDBC, 96 percent of the water consumed within the basin in 2003-2004 was within the cap.[9]

With a lid on extractions, new water demands in the Murray-Darling basin are met primarily through conservation, efficiency improvements, and water trading. Most of the early buying and selling of water entitlements has occurred within states, but the MDBC is now piloting a program in the southern portion of the basin to allow permanent water trades across state boundaries.[10] The initial two-year review of this scheme found that it had enabled 51 trades collectively worth about Aus$10 million, which had transferred nearly 10 million cubic meters of water between states.[11] With virtually all of the traded water going to higher-value uses, water marketing is boosting the basin's money economy. Indeed, a 1999 study projected a doubling of the basin's economy over 25 years with the cap and water reforms in place.[12]

There is, however, an important hitch: the cap was pegged to a level of withdrawals that had allowed serious degradation of the river's health. So while it may prevent further deterioration, the cap is not sufficiently stringent to revitalize the river.[13] More recently, the cap has been augmented by the Living Murray initiative, a major effort to return more flows to the river.[14]

Limits on groundwater pumping to stem the depletion of underground aquifers are a high priority as well. In the U.S. state of Texas, a cap on groundwater use is now in place for the Edwards Aquifer, a major source of irrigation water in south-central Texas and of drinking water for the city of San Antonio.[15] By the early 1990s, heavy pumping from the aquifer had substantially reduced flows in San Marcos and Comal Springs, which harbor seven species listed under the federal Endangered Species Act, including the Texas Blind Salamander and the Fountain Darter. The Sierra Club and others filed a lawsuit under the Act to limit pumping so as to sustain flows in the springs. In response, the Texas legislature established the Edwards Aquifer Authority in 1993 and set a 555.3 million cubic meter cap on annual pumping from the aquifer through 2007 and a more stringent cap of 493.6 million cubic meters by 2008.[16] In addition, the Authority is to have enforceable procedures in place by 2012 to ensure continuous minimum flows for the two springs.

As in the Murray-Darling basin, the cap on withdrawals from the Edwards Aquifer has fostered an active water market. Most of the trades, which include both permanent sales and temporary leases of water, involve irrigators selling water to San Antonio. To date, irrigators have traded some 185.1 million cubic meters of water per year to urban users.[17] The cap has also encouraged more conservation in San Antonio, where per capita domestic use is now considerably lower than in most Texas cities.[18]

The institution of the Edwards Aquifer cap represents a marked departure from Texas' long-standing "rule of capture" —sometimes called the "rule of the biggest pump"—which essentially allows landowners to withdraw as much ground-water from beneath their land as they want to, as long as they put it to some beneficial use. Harm to neighbors or the environment does not constrain pumping rights under the rule of capture. This antiquated rule still governs much of the groundwater in Texas, but perhaps experience with the Edwards Aquifer will encourage broader policy reform.

Effective pricing of water is an underused tool for spurring more-efficient water use in agriculture, industry, and home environments. Many utilities and irrigation authorities still charge a flat fee for water; some even charge lower unit prices the more a customer consumes. By contrast, tiered pricing, in which the unit price of water increases in stair-step fashion along with the volume used, can encourage conservation. Suppliers can make the first tier a "lifeline" quantity that is priced very low (even at zero) to ensure that poor households receive enough water to meet their basic needs. The higher prices paid by profligate users can help subsidize these lifeline supplies, building equity into the pricing scheme. As sensible as these conservation-oriented rate structures are, they are still greatly underused in rich and poor countries alike. A 2002 study of 300 Indian cities found that only 13 percent use such tiered rate structures.[19]

Price incentives are also critical for irrigators, and can even exist in tandem with some degree of subsidy. One idea being tried on California farms, for instance, is a three-tiered water rate structure under which irrigation districts pay the same rates they were previously paying, plus a fixed surcharge, for up to 80 percent of their contracted water volume.[20] If their usage

SIDEBAR 6

Twelve Priorities for Updating Water Policies

- Make watershed protection an integral part of drinking water supply and rural development.
- Inventory the health status of freshwater ecosystems and set ecological goals.
- Establish caps on river modification, groundwater pumping, nutrient discharges, and watershed degradation in order to safeguard ecosystem services.
- Call on water authorities to operate dams so that river flows better resemble the natural flow regime.
- Encourage water trading and payments for ecosystem services that help to achieve ecological goals equitably and efficiently.
- Reduce irrigation subsidies and institute tiered water pricing structures that encourage conservation and efficiency.
- Establish conservation and efficiency standards for municipal, industrial, landscape, and irrigation water use.
- Boost investments in affordable irrigation technologies and methods to enable poor farmers to raise their land and water productivity.
- Extend training and scientific advice to lift rainfed crop production in poor regions.
- Increase the monitoring and surveillance of stream flows and watershed conditions.
- Educate citizens on how personal choices, from diets to outdoor landscapes, can reduce their individual claim on freshwater ecosystems.
- Ensure that decision making is inclusive, transparent, and accountable to the public, and encourage citizen involvement in water management.

---

equals 80–90 percent of their contract allotment, however, they pay substantially more per cubic meter for that additional 10 percent, and if their usage goes as high as 90–100 percent of their contract amount, they pay even more for that last 10 percent increment—in some cases, nearly three times more than the base level. Irrigators therefore have an incentive to reduce their water use in order to avoid paying the higher unit costs for the last 10 or 20 percent of their contracted allotment. One irrigation district that implemented a similar pricing scheme in the late 1980s saw its average water use decline by 19 percent within a few years.

Based on these and other measures discussed earlier, twelve priorities emerge for updating water policies so they work to protect, not degrade, freshwater ecosystems. (See Sidebar 6, page 59.) Among them is a call for stepped-up monitoring and surveillance of river flows, groundwater levels, and watershed health. The global network of streamflow gauges and hydrological monitoring stations has deteriorated markedly over the last couple of decades. Society's ability to respond to the hydrological changes that are occurring requires reliable information that only good monitoring can provide. Managing rivers to better match their natural flow patterns, for instance, requires that scientists have enough data on water use and river levels throughout a watershed to pinpoint opportunities to restore ecologically important flows.

Leadership, commitment, and citizen involvement are the driving forces behind many of the most innovative and successful water projects and policy reforms. Most of these efforts began with a small number of committed individuals, organizations, water managers, or political leaders who decided to buck the odds and push for a different approach. Others now need to augment their efforts. Old approaches and entrenched ways die hard. But the benefits of working constructively with nature's water cycle, rather than further disrupting it, are now too compelling to ignore.

Endnotes

Introduction

1. Christopher Pala, "To Save a Vanishing Sea," *Science*, 18 February 2005, pp. 1032–34; Philip Micklin, "Touring the Aral: Visit to an Ecological Disaster Zone," *Soviet Geography*, February 1991; author's visit to Aral Sea region, March 1995.

2. Fluxes are approximate; estimates of global annual precipitation over land, for example, range from 107,000–119,000 cubic kilometers, according to United Nations, *Water for People, Water for Life: The United Nations World Water Development Report* (Paris: UNESCO Publishing and Berghahn Books, 2003), p. 77. Figure 1 adapted from Robert B. Jackson et al., "Water in a Changing World," *Issues in Ecology* (Ecological Society of America), No. 9, Spring 2001 and from Sandra L. Postel, Gretchen C. Daily, and Paul R. Ehrlich, "Human Appropriation of Renewable Fresh Water," *Science*, 9 February 1996, pp. 785–88.

3. For a description of early Nile Valley agriculture, see Sandra Postel, *Pillar of Sand: Can the Irrigation Miracle Last?* (New York: W.W. Norton & Company, 1999).

4. Jared Diamond, "Lessons from Environmental Collapses of Past Societies," Fourth Annual John H. Chafee Memorial Lecture on Science and the Environment (Washington, DC: National Council for Science and the Environment, 2004); Jared Diamond, *Collapse: How Societies Choose to Fail or Survive* (New York: Viking Books, 2004).

5. Sumerians, Harappan, and Hohokam from Postel, op. cit. note 3.

6. David Dudgeon, "Large-Scale Hydrological Changes in Tropical Asia: Prospects for Riverine Biodiversity," *BioScience*, September 2000, pp. 793–806.

7. Author's calculation based on population increase from medium variant estimates from United Nations Population Division, *World Population Prospects: The 2004 Revision*, at esa.un.org/unpp, viewed 18 May 2005 and current global average dietary water requirement of 1,200 cubic meters per person from Malin Falkenmark and Johan Rockström, *Balancing Water for Humans and Nature: The New Approach in Ecohydrology* (London: Earthscan, 2004). However, dietary trends could significantly increase or decrease this average by 2030.

8. U.N. Millennium Development Goals, at www.un.org/millenniumgoals/, viewed 18 May 2005; Swedish Water House and Millennium Project, "Investing in the Future: Water's Role in Achieving the Millennium Development Goals," Policy Brief No. 1 (Stockholm: undated).

9. Mike Dombeck, "The Forgotten Forest Product: Water," *New York Times*, 3 January 2003.

Assessing the Damage—and How We Got Where We Are

1. Figure of 6.4 billion from United Nations Population Division, *World Population Prospects: The 2004 Revision*, at esa.un.org/unpp, viewed 18 May 2005; $55 trillion from Erik Assadourian, "Global Economy Continues to Grow," in Worldwatch Institute, *Vital Signs 2005* (New York: W.W. Norton & Company, 2005), pp. 44–45.

2. World Commission on Dams (WCD), *Dams and Development* (London: Earthscan, 2000).

3. Sandra Postel, *Pillar of Sand: Can the Irrigation Miracle Last?* (New York: W.W. Norton & Company, 1999.

4. Ibid.

5. U.N. Food and Agriculture Organization, FAOSTAT, electronic database, at faostat.fao.org.

6. Sidebar 2 is based on the following sources: Sandra Postel and Brian Richter, *Rivers for Life: Managing Water for People and Nature* (Washington DC: Island Press, 2003); Sandra Postel and Stephen Carpenter, "Freshwater Ecosystem Services," in Gretchen C. Daily, ed., *Nature's Services: Societal Dependence on Natural Ecosystems* (Washington, DC: Island Press, 1997), pp. 195–214. Watershed conversion figures from Carmen Revenga, Siobhan Murray, Janet Abramovitz, and Allen Hammond, *Watersheds of the World: Ecological Value and Vulnerability* (Washington, DC: World Resources Institute (WRI), 1998); wetlands from Rudy Rabbinge and Prem S. Bindraban, "Poverty, Agriculture, and Biodiversity," in John A. Riggs, ed., *Conserving Biodiversity* (Washington, DC: The Aspen Institute, 2005), pp. 65–77; number of dams from WCD, op. cit. note 2. The WCD defines a large dam as being at least 15 meters high or between 5 and 15 meters high and with a reservoir volume of more than 3 million cubic meters. Dam effects from Christer Nilsson et al., "Fragmentation and Flow Regulation of the World's Large River Systems," *Science*, 15 April 2005, pp. 405–08 and from Matts Dynesius and Christer Nilsson, "Fragmentation and Flow Regulation of River Systems in the Northern Third of the World," *Science*, vol. 266 (1994), pp. 753–62; interception of river flows from Charles Vörösmarty and Dork Sahagian, "Anthropogenic Disturbance of the Terrestrial Water Cycle," *BioScience*, September 2000, pp. 753–65. Percentages calculated by author assuming 40,000 cubic kilometers per year of global runoff. Figure of 15 percent from Nilsson et al., op. cit. this note; 100 billion tons from James P.M. Syvitski et al., "Impact of Humans on the Flux of Terrestrial Sediment to the Global Coastal Ocean," *Science*, 15 April 2005, pp. 376–80; nitrogen from International Fertilizer Industry Association, "Nitrogen Fertilizer Nutrient Consumption," electronic database, at www.fertilizer.org/ifa/statistics/indicators/tablen.asp, viewed 28 January 2005; 7 billion tons, 35 percent, and 10 warmest years from Janet L. Sawin, "Climate Change Indicators on the Rise," in Worldwatch Institute, *Vital Signs 2005* (New York: W.W. Norton & Company, 2005), pp. 40–41; 20 percent from Peter B. Moyle and Robert

A. Leidy, "Loss of Biodiversity in Aquatic Ecosystems: Evidence from Fish Faunas," in P.L. Fiedler and S.K. Jain, eds., *Conservation Biology: The Theory and Practice of Nature Conservation, Preservation, and Management* (New York: Chapman and Hall, 1992); 6.4 billion from United Nations Population Division, op. cit. note 1; water use from Postel, op. cit. note 3; wood and energy consumption from Gary Gardner, Erik Assadourian, and Radhika Sarin, "The State of Consumption Today," in Worldwatch Institute, *State of the World 2004* (New York: W.W. Norton & Company, 2004), p. 17.

7. Historical flows from Philip Micklin, "Managing Transnational Waters of the Aral Sea Basin: A Geographical Perspective," prepared for the conference "Agricultural Development in Central Asia, Between Russia and the Middle East," University of Washington, Seattle, 20–22 November 1998.

8. Figure 2 is based on the following sources: 1990–2003 from Philip P. Micklin, Western Michigan University, Kalamazoo, e-mail to author, February 2005. Other years calculated by Philip P. Micklin, based on data from A. Ye. Asarin and V.N. Bortnik (1926–1985) and other sources (1986–1989), as cited in Peter H. Gleick, ed., *Water in Crisis* (New York: Oxford University Press, 1993), p. 314.

9. Small Aral from Christopher Pala, "To Save a Vanishing Sea," *Science*, 18 February 2005, pp. 1032–34; 1987 split from Micklin, op. cit. note 7.

10. Wang Shucheng, "Water Resources Management of the Yellow River and Sustainable Water Development in China," *Water Policy*, vol. 5 (2003), pp. 305–312.

11. Length and duration of dryness from Ma Jun, *China's Water Crisis* (Norwalk, CT: EastBridge, 2004) (originally published in Chinese in 1999); $1.6 billion from "China's Yellow River Flows Freely for 5 Consecutive Years," *Xinhua Economic News Service*, 29 December 2004.

12. Sandra Postel, "Where Have all the Rivers Gone?" *World Watch*, May/June 1995, pp. 9–19.

13. Jason P. Ericson et al., "Effective Sea-Level Rise in Deltas: Sources of Change and Human-Dimension Implications," *Global and Planetary Change*, in press (2005).

14. Figure 3 derived from the following sources: 1904–1949 from U.S. Geological Survey, "Calendar Year Streamflow Statistics for the Nation: USGS 09521000 Colorado River at Yuma, AZ," at nwis.waterdata.usgs.gov/nwis/annual/calendar_year/?site_no=09521000, viewed 28 January 2005; 1950–2004 is flow at Southerly International Boundary, provided by Kenneth Rakestraw, International Boundary and Water Commission, personal communication to author, 22 December 2004.

15. Charles Bergman, *Red Delta: Fighting for Life at the End of the Colorado River*

(Golden, CO: Fulcrum Publishing, 2002); M.S. Galindo-Bect et al., "Analysis of Penaeid Shrimp Catch in the Northern Gulf of California in Relation to Colorado River Discharge," *Fishery Biology*, vol. 98 (2000), pp. 222–25.

16. Postel, op. cit. note 3.

17. "International Conference on Indus Delta Eco Region Calls for Release of Minimum Environmental Flows Downstream Kotri," *Global News Wire* (Pakistan Press International Information Services Limited), 7 October 2004.

18. Erik Eckholm, "A River Diverted, the Sea Rushes In," *New York Times*, 22 April 2003.

19. Ibid.

20. IUCN-The World Conservation Union, *Value: Counting Ecosystems as Water Infrastructure* (Gland, Switzerland: IUCN, 2004), p. 22.

21. "Fishermen to Stage Sit-in at Sujawal Bridge Against Water Shortage in River Indus Downstream Kotri," The *Pakistan Newswire* (Pakistan Press International), 24 July 2004.

22. N. LeRoy Poff et al., "The Natural Flow Regime," *BioScience*, December 1997, pp. 769–84.

23. Postel and Richter, op. cit. note 6.

24. Figure 4 from Postel and Richter, op. cit. note 6.

25. David L. Galat and Robin Lipkin, "Restoring Ecological Integrity of Great Rivers: Historical Hydrographs Aid in Defining Reference Conditions for the Missouri River," *Hydrobiologia*, vol. 422/423 (2000), pp. 29–48.

26. National Research Council, *The Missouri River Ecosystem: Exploring the Prospects for Recovery* (Washington, DC: National Academy Press, 2002).

27. Bruce Stein, Lynn S. Kutner, and Jonathan S. Adams, eds., *Precious Heritage: The Status of Biodiversity in the United States* (New York: Oxford University Press, 2000).

28. Charles Lydeard et al., "The Global Decline of Nonmarine Mollusks," *BioScience*, April 2004, pp. 321–30.

29. Anthony Ricciardi and Joseph B. Rasmussen, "Extinction Rates of North American Freshwater Fauna," *Conservation Biology*, vol. 13 (1999), pp. 1220–22.

30. Michel Meybeck, "Global Analysis of River Systems: From Earth System Controls to Anthropocene Syndromes," *Phil. Trans. R. Soc. Lond. B*, vol. 358 (2003), pp. 1935–55.

31. Jonathan J. Cole et al., "Nitrogen Loading of Rivers as a Human-Driven Process," in M.J. McDonnell and S.T.A. Pickett., eds., *Humans as Components of Ecosystems: The Ecology of Subtle Human Effects and Populated Areas* (New York: Springer-Verlag, 1993).

32. Pamela A. Green et al., "Pre-industrial and Contemporary Fluxes of Nitrogen through Rivers: A Global Assessment Based on Typology," *Biogeochemistry*, vol. 68 (2004), pp. 71–105.

33. Robert Howarth et al., "Nutrient Pollution of Coastal Rivers, Bays, and Seas," *Issues in Ecology* (Ecological Society of America), Fall 2000.

34. United Nations Environment Programme, *Global Environment Outlook 3* (Nairobi: 2002).

35. Robert J. Diaz, "Overview of Hypoxia around the World," *Journal of Environmental Quality*, vol. 30 (2001), pp. 275–81.

36. Gulf of Mexico from N. Rabalais, R. Turner, and D. Scavia, "Beyond Science into Policy: Gulf of Mexico Hypoxia and the Mississippi River," *BioScience*, February 2002, pp. 129–42 and from Howarth et al., op. cit. note 33; East China Sea from Li Daoji and Dag Daler, "Ocean Pollution from Land-based Sources: East China Sea, China," *Ambio*, vol. 33 (2004), pp. 107–13; Baltic Sea from Rabalais, Turner, and Scavia, op. cit. this note and from R. Elmgren, "Understanding Human Impact on the Baltic Ecosystem: Changing Views in the Recent Decades," *Ambio*, vol. 30 (2001), pp. 222–31.

37. Diaz, op. cit. note 35.

38. Gerald Niemi et al., "Rationale for a New Generation of Indicators for Coastal Waters," *Environmental Health Perspectives*, vol. 112 (2004), pp. 979–86.

39. Figure 5 from International Fertilizer Industry Association (IFIA), op. cit. note 6. Here, category of Southeast and East Asia includes IFIA groupings of East Asia, Northeast Asia, and Southeast Asia.

40. Ibid.

41. Intergovernmental Panel on Climate Change Working Group II, *Summary for Policymakers—Climate Change 2001: Impacts, Adaptation and Vulnerability* (Geneva, Switzerland: approved February 2001).

42. United Nations University-Environment and Human Security, "Two Billion People Vulnerable to Floods by 2050; Number Expected to Double or More in Two Generations due to Climate Change, Deforestation, Rising Seas, Population Growth," press release (Tokyo: 13 June 2004).

43. Petra Döll, "Impact of Climate Change and Variability on Irrigation

Requirements: A Global Perspective," *Climatic Change*, August 2002, pp. 269–293.

44. Fulu Tao et al., "Terrestrial Water Cycle and the Impact of Climate Change," *Ambio*, June 2003, pp. 295–301.

45. Robert F. Service, "As the West Goes Dry," *Science*, 20 February 2004, pp. 1124–27.

46. "Glacier Meltdown," *New Scientist*, 8 May 2004, p. 7.

47. Juan Forero, "As Andean Glaciers Shrink, Water Worries Grow," *New York Times*, 24 November 2002.

48. Gallaire quote from ibid.

Healthy Watersheds for Safe Drinking Water

1.   Nigel Dudley and Sue Stolton, eds., *Running Pure: The Importance of Forest Protected Areas to Drinking Water* (Gland, Switzerland: The World Bank/WWF Alliance for Forest Conservation and Sustainable Use, 2003), p. 40.

2.   Juan D. Quintero, Lead Environmental Specialist for Latin America and Caribbean Region, Environmentally and Socially Sustainable Development Department, World Bank, Washington, DC, personal communication with author, 27 May 2004.

3.   International Consortium of Investigative Journalists, *The Water Barons* (Washington, DC: Public Integrity Books, 2003), p. 108.

4.   Sarah Garland, "Keeping it Public in Bogotá," *NACLA Report on the Americas,* July-August 2004.

5.   Quintero, op. cit. note 2.

6.   R. Hirji and H.O. Ibrekk, "Environmental and Water Resources Management," *World Bank Environment Strategy Papers*, No. 2 (Washington, DC: World Bank, 2001).

7.   Table 1 adapted from Sandra L. Postel and Barton H. Thompson, Jr., "Watershed Protection: Capturing the Benefits of Nature's Water Supply Services," *Natural Resources Forum*, May 2005, pp. 98–108 and from Walter V. Reid, "Capturing the Value of Ecosystem Services to Protect Biodiversity," in V.C. Hollowell, ed., *Managing Human Dominated Ecosystems* (St. Louis: Missouri Botanical Garden, 2001).

8.   National Research Council, *Watershed Management for Potable Water Supply: Assessing the New York City Strategy* (Washington, DC: National Academy Press, 2000).

9. Ibid.

10. Figure of $1 billion from Christopher Ward, "From Commissioner Chris Ward," *Around the Watershed* (New York City Department of Environmental Protection), Winter 2004.

11. Reid, op. cit. note 7.

12. Dudley and Stolton, op. cit. note 1.

13. Sidebar 3 is based on the following sources: Marta Echavarria, "Financing Watershed Conservation: The FONAG Water Fund in Quito, Ecuador," in S. Pagiola, J. Bishop, and N. Landell-Mills, eds., *Selling Forest Environmental Services: Market-based Mechanisms for Conservation and Development* (London: Earthscan, 2002); $6,000 from "FONAG: Quito's Water Fund: A Municipal Commitment to Protect the Watershed," at www.unep.org/GC/GCSS-VIII/USA-IWRM-2.pdf, viewed 1 March 2005; 2004 funding from Pablo Lloret, "A Trust Fund as a Financial Instrument for Water Protection and Conservation: The Case of the Environmental Water Fund in Quito, Ecuador," prepared for Conference on Water for Food and Ecosystems: Make it Happen!, at www.fao.org/ag/wfe2005/docs/Fonag_Ecuador_en.pdf, and from "FONAG: Quito's Water Fund...," op. cit. this note.

14. Dudley and Stolton, op. cit. note 1.

15. Brian W. van Wilgen, Richard M. Cowling, and Chris J. Burgers, "Valuation of Ecosystem Services: A Case Study from South African Fynbos Ecosystems," *BioScience*, March 1996, pp. 184–89.

16. Government of South Africa, Working for Water Programme Web site, at www-dwaf.pwv.gov.za/wfw.

17. Guy Preston, "Invasive Alien Plants and Protected Areas in South Africa," paper presented at the World Parks Congress, Durban, South Africa, 13 September 2003.

18. Sandra Postel and Amy Vickers, "Boosting Water Productivity," in Worldwatch Institute, *State of the World 2004* (New York: W.W. Norton & Company, 2004), pp. 46–65.

19. Ibid.

20. Amy Vickers, *Handbook of Water Use and Conservation: Homes, Landscapes, Businesses, Industries, and Farms* (Amherst, MA: WaterPlow Press, 2001).

21. Postel and Vickers, op. cit. note 18.

22. Sidebar 4 is based on the following sources: water demand in 2004 from Jonathan Yeo, Massachusetts Water Resources Authority (MWRA), e-mail to author, 18 January 2005. Figure 6 data from the MWRA as provided by Eileen Simonson, The Water Supply Citizens Advisory Committee to the MWRA, Hadley, Massachusetts, e-mail to author, March 2005; MWRA Web site, at www.mwra.com/04water/html/wsupdate.htm, viewed 28 February 2005; capital cost savings from Amy Vickers & Associates, Inc., "Final Report: Water Conservation Planning USA Case Studies Project," prepared for Environment Agency, Demand Management Centre, United Kingdom (Worthing, West Sussex: June 1996); lawn watering from Simonson, op. cit. this note; "The Water Supply Citizens Advisory Committee to the MWRA: Government-Supported Public Participation," at www.mwra.com/02org/html/wscac.htm, viewed 28 February 2005.

23. Pauline Boerma, "Watershed Management: A Review of the World Bank Portfolio (1990–1999)," (Washington, DC: Rural Development Department, World Bank, 2000).

24. World Bank, "Implementation Completion Report (CPL-31600; SCPD-3160S) on a loan in the amount of US$ Million 33.0 to the Federative Republic of Brazil for Land Management II-Santa Catarina Project (Loan 3160-BR)" (Washington, DC: 2000); crop productivity from G. Lituma, M.I. Braga, and A. Soler, "Scaling Up Watershed Management Projects: The Experience of Southern Brazil," presentation at World Bank Water Week, World Bank, Washington, DC, March 2003.

25. Lauro Bassi, "Valuation of Land Use and Management Impacts on Water Resources in the Lajeado São José Micro-Watershed, Chapecó, Santa Catarina State, Brazil," prepared for e-Workshop on *Land-Water Linkages in Rural Watersheds: Case Study Series*, FAO, Rome, 2002.

26. Ahmad Hijjawi, "Investment in Water Efficiency Will Improve Household Water Supply—Report," (Amman: Jordan Information Center, posted 31 March 2005).

27. Vickers, op. cit. note 20.

28. Kevin Hurley, "Prozac Seeping into Water Supplies," *The Scotsman*, 9 August 2004.

29. Dana W. Kolpin et al., "Pharmaceuticals, Hormones, and Other Organic Wastewater Contaminants in U.S. Streams, 1999–2000: A National Reconnaissance," *Environmental Science and Technology*, 15 March 2002, pp. 1202–11.

30. Andrea B. Kirk et al., "Perchlorate and Iodide in Dairy and Breast Milk," *Environmental Science and Technology*, vol. 39 (2005), pp. 2011–17.

31. Figure of 11 million from "Study Shows Perchlorate in Water Supply of Millions of Americans," *U.S. Water News*, March 2005. To date federal officials have not established a national drinking water standard for perchlorate.

32. U.S. Environmental Protection Agency, "National Perchlorate Detections as of September 23, 2004," at www.epa.gov/fedfac/documents/perchlorate _map/nationalmap.htm.

33. Rachel Carson, *Silent Spring* (New York: Houghton-Mifflin, 1962); Theo Colborn, Dianne Dumanoski, and John Peterson Myers, *Our Stolen Future* (New York: Penguin Books, 1996).

34. Colborn, Dumanoski, and Myers, op. cit. note 33, p. 38.

Food Security With Ecosystem Security

1. D. Renault and W.W. Wallender, "Nutritional Water Productivity and Diets," *Agricultural Water Management*, vol. 45 (2000), pp. 275–96.

2. A.Y. Hoekstra and P.Q. Hung, "Virtual Water Trade: A Quantification of Virtual Water Flows Between Nations in Relation to International Crop Trade," *Value of Water Research Report Series*, No. 11 (Delft, The Netherlands: IHE Delft, September 2002). Water requirements for grain can rise to 5,000 liters per kilogram in very dry climates.

3. Sandra Postel and Amy Vickers, "Boosting Water Productivity," in Worldwatch Institute, *State of the World 2004* (New York: W.W. Norton & Company, 2004), pp. 46–65.

4. Sandra Postel, *Pillar of Sand: Can the Irrigation Miracle Last?* (New York: W.W. Norton & Company, 1999).

5. Tushaar Shah et al., "Sustaining Asia's Groundwater Boom: An Overview of Issues and Evidence," *Natural Resources Forum*, May 2003, pp. 130–41.

6. Sandra Postel, "Securing Water for People, Crops, and Ecosystems: New Mindset, and New Priorities," *Natural Resources Forum*, May 2003, pp. 89–98.

7. U.N. Food and Agriculture Organization (FAO), "Hunger Costs Millions of Lives and Billions of Dollars–FAO Hunger Report," press release (Rome: 8 December 2004).

8. Postel, op. cit. note 4.

9. Ibid.

10. "Irrigation Options," WCD Thematic Review IV.2, as reported in World Commission on Dams, *Dams and Development* (London: Earthscan, 2000).

11. Nile River from P.M. Chesworth, "The History of Water Use in the Sudan and Egypt," in P.P. Howell and J.A. Allan, eds., *The Nile: Sharing a Scarce Resource* (Cambridge: Cambridge University Press, 1994), pp. 65–79.

12. Calculation based on withdrawal estimates in William J. Cosgrove and Frank R. Rijsberman, *World Water Vision: Making Water Everybody's Business* (London: Earthscan, 2000).

13. For a fuller discussion of options to improve irrigation water productivity, see Postel, op. cit. note 4.

14. Postel and Vickers, op. cit. note 3.

15. Cost figure from Shilp Verma, Stanzin Tsephal, and Tony Jose, "Pepsee Systems: Grassroots Innovation under Groundwater Stress," *Water Policy*, vol. 6 (2004), pp. 303–18.

16. Paul Polak, President, International Development Enterprises, personal communication with author, 28 December 2004.

17. Prem Bindraban and Huib Hengsdijk, "Water-Saving in Rice-based Ecosystems," submission to the FAO conference on Water for Food and Ecosystems, The Hague, The Netherlands, 31 January–4 February 2005.

18. L.C. Guerra et al., *Producing More Rice with Less Water from Irrigated Systems* (Colombo, Sri Lanka: International Water Management Institute (IWMI), 1998), p. 11; R. Barker, Y.H. Li, and T.P. Tuong, eds., *Water-Saving Irrigation for Rice, Proceedings of an International Workshop held in Wuhan, China, 23–25 March 2001* (Colombo, Sri Lanka: IWMI, 2001).

19. Indonesia from Bindraban and Hengsdijk, op. cit. note 17.

20. Cornell International Institute for Food, Agriculture and Development, "The System of Rice Intensification," at ciifad.cornell.edu/sri, viewed 1 May 2005. See also Association Tefy Saina Web site, at www.tefysaina.org.

21. FAO, op. cit. note 7.

22. Sandra L. Postel, "Water for Food Production: Will There Be Enough in 2025? *BioScience*, August 1998, pp. 629–37.

23. Postel, op. cit. note 4.

24. Malin Falkenmark and Johan Rockström, *Balancing Water for Humans and Nature: The New Approach in Ecohydrology* (London: Earthscan, 2004).

25. Syria from Jim S. Wallace and Peter J. Gregory, "Water Resources and their Use in Food Production Systems," *Aquatic Sciences*, vol. 64 (2002), pp. 363–75; China from Falkenmark and Rockström, op. cit. note 24.

26. Falkenmark and Rockström, op. cit. note 24; Gordon Conway and Gary Toenniessen, "Science for African Food Security," *Science*, 21 February 2003, pp. 1187–88.

27. See Wallace and Gregory, op. cit. note 25, for a concise menu of water productivity options.

28. Johan Rockström, "Water for Food and Nature in Drought-Prone Tropics: Vapour Shift in Rain-fed Agriculture," *Phil. Trans. R. Soc. Lond. B*. vol. 358 (2003), pp. 1997–2009.

29. Ibid.

30. Adrian Wood and Alan Dixon, "Sustainable Wetland Management and Food Security: The Role of Integrated Multiple Use Regimes in the Upper-Baro Basin, South-West Ethiopia," submission to the FAO conference on Water for Food and Ecosystems, The Hague, The Netherlands, 31 January–4 February 2005.

31. Figure of 250 years from Alan B. Dixon and Adrian P. Wood, "Wetland Cultivation and Hydrological Management in Eastern Africa: Matching Community and Hydrological Needs through Sustainable Wetland Use," *Natural Resources Forum*, May 2003, pp. 117–29.

32. Lucy Emerton and Elroy Bos, *Value: Counting Ecosystems as Water Infrastructure* (Gland, Switzerland: IUCN, 2004), p. 21.

33. Sidebar 5 from Edward B. Barbier and Julian R. Thompson, "The Value of Water: Floodplain Versus Large-scale Irrigation Benefits in Northern Nigeria," *Ambio*, vol. 27 (1998), pp. 434–40.

34. Dietary water requirements and Figure 7 from Renault and Wallender, op. cit. note 1.

35. Ibid.

36. The medium variant U.S. population projection for 2025 is 350 million, according to United Nations Population Division, *World Population Prospects: The 2004 Revision*, at esa.un.org/unpp, viewed 18 May 2005. The 200 million figure assumes an annual average dietary water requirement of 1,242 cubic meters per person, derived by reducing current annual U.S. dietary water use of 1,971 cubic meters per person (from Renault and Wallender, op. cit. note 1) by 37 percent.

Reducing Risks, Preserving Resilience

1. An estimated 3,300 and 1,500 Haitians were lost in the storms in May and September, respectively, according to Lesly C. Hallman, "Death Toll in Haiti Continues to Rise," news release (Washington, DC: American Red Cross, 28 September 2004).

2. Deborah Sontag and Lydia Polgreen, "Storm-Battered Haiti's Endless Crises Deepen," *New York Times*, 16 October 2004.

3. Differing impact on Haiti and Puerto Rico from T. Mitchell Aide and H. Ricardo Grau, "Globalization, Migration, and Latin American Ecosystems," *Science*, 24 September 2004, pp. 1915–16, and from Hallman, op. cit. note 1, who reports that seven died from the September event in Puerto Rico.

4. Figure of 273,000 from International Federation of Red Cross and Red Crescent Societies, "Asia: Earthquake and Tsunamis," Fact Sheet No. 8 (Geneva: 24 March 2005). For Indonesia and India, the number of dead includes those listed as missing.

5. "EarthTalk: Is it True that Coastal Development Contributed to Greater Loss of Life from the Tsunami?" *E/The Environment Magazine*, 11 January 2005.

6. David Fogarty, "Tsunami-Hit Nations Look to Save Mangroves," *Reuters*, 17 January 2005.

7. Munich Re Group, *Annual Review: Natural Catastrophes 2004* (Munich: December 2004).

8. J.T. Houghton et al., eds., *Climate Change 2001: The Scientific Basis, Contribution of Working Group I to the Third Assessment Report of the Intergovernmental Panel on Climate Change* (Cambridge, UK: Cambridge University Press, 2001); Janet L. Sawin, "Severe Weather Events on the Rise," in Worldwatch Institute, *Vital Signs 2003* (New York: W.W. Norton & Company, 2003), pp. 92–93.

9. Stefano Pagiola, Environment Department, World Bank, e-mail to author, 4 May 2004.

10. Janet N. Abramovitz, *Unnatural Disasters*, Worldwatch Paper 158 (Washington, DC: Worldwatch Institute, 2001).

11. European Commission, "Flood Protection: Commission Proposes Concerted EU Action," press release (Brussels: 12 July 2004).

12. Population figure from European Commission, "Towards a European Action Program on Flood Risk Management," at europa.eu.int/comm/environment/water/flood_risk/index.htm, viewed 29 March 2005.

13. Abramovitz, op. cit. note 10.

14. European Commission, op. cit. note 12.

15. Karen F. Schmidt, "A True-Blue Vision for the Danube," *Science*, 16 November 2001, pp. 1444–47.

16. World Wide Fund for Nature, Living Waters Program-Europe, "A Green Corridor for the Danube," as described in Sandra Postel and Brian Richter,

Rivers for Life: Managing Water for People and Nature (Washington, DC: Island Press, 2003).

17. Schmidt, op. cit. note 15.

18. Abramovitz, op. cit. note 10.

19. E. Rykiel, "Ecosystem Science for the Twenty-First Century," *BioScience*, October 1997, pp. 705–08; $19 billion from Abramovitz, op. cit. note 10.

20. National Research Council, *Valuing Ecosystem Services: Toward Better Environmental Decision-Making* (Washington, DC: The National Academy Press, 2005), p. 170.

21. Ibid.

22. Gretchen C. Daily and Katherine Ellison, *The New Economy of Nature: The Quest to Make Conservation Profitable* (Washington, DC: Island Press, 2002).

Bringing Water Policies Into the 21st Century

1. South African National Water Act No. 36 of 1998, *Government Gazette*, Vol. 398, No. 19182 (Cape Town: 26 August 1998).

2. South African National Water Act No. 36 of 1998, Part 3: "The Reserve," and Appendix 1: "Fundamental Principles and Objectives for a New Water Law in South Africa," *Government Gazette*, Vol. 398, No. 19182 (Cape Town: 26 August 1998).

3. International Conference on Freshwater, *Water—A Key to Sustainable Development: Recommendations for Action,* Bonn, Germany, 3–7 December 2001.

4. Millennium Ecosystem Assessment, *Ecosystems and Human Well-being: Synthesis* (Washington, DC: Island Press, 2005).

5. Sandra Postel and Brian Richter, *Rivers for Life: Managing Water for People and Nature* (Washington DC: Island Press, 2003).

6. Table 2 is based on the following sources: Murray-Darling from Don J. Blackmore, "The Murray-Darling Basin Cap on Diversions—Policy and Practice for the New Millennium," *National Water*, 15–16 June 1999, pp. 1–12; Great Lakes from Postel and Richter, op. cit. note 5 and from Council of Great Lakes Governors, "Great Lakes Water Management Initiative: Draft Annex 2001 Implementing Agreements" at www.cglg.org/projects/water/annex2001imple menting.asp, viewed 18 May 2005; European Union Directive from Postel and Richter, op. cit. note 5; Ipswich from Massachusetts Department of Environmental Protection, "State Strikes Balance with Water Withdrawal Permits for

Ipswich River Basin Communities," press release (Boston: 20 May 2003) (Note: permitting changes are now being contested in the courts); Yellow River from "China's Yellow River Flows Freely for 5 Consecutive Years," *Xinhua Economic News Service*, 29 December 2004 and from Wang Shucheng, "Water Resources Management of the Yellow River and Sustainable Water Development in China," *Water Policy*, vol. 5 (2003), pp. 305–312; Edwards Aquifer from Mary Kelly, "A Powerful Thirst: Water Marketing in Texas" (Austin, Texas: Environmental Defense, 2004); Pamlico Estuary from North Carolina Department of Environment and Natural Resources, "Nonpoint Source Management Program, Tar-Pamlico Nutrient Strategy," at h2o.enr.state.nc.us/nps/tarpam.htm, viewed 27 February 2004, and from Environomics, "A Summary of U.S. Effluent Trading and Offset Projects," report prepared for U.S. Environmental Protection Agency, Office of Water (Bethesda, MD: November 1999).

7. Blackmore, op. cit. note 6.

8. Impact in 2003 from "Drying Out," *The Economist*, 12 July 2003, p. 38.

9. Murray-Darling Basin Commission (MDBC), *Annual Report 2003–04* (Canberra, Australian Capital Territory (ACT): 2004). The actual diversions allowed under the cap vary from year to year depending on climatic and hydrologic conditions, but are pegged to 1993/94 withdrawal levels as described in *The Cap* (Canberra, ACT: MDBC, 2004).

10. MDBC, "Pilot Interstate Water Trading Project," at www.mdbc.gov.au/naturalresources/watertrade/pilot_watertrade.htm.

11. Mike Young et al., *Inter-State Water Trading: A Two Year Review* (Canberra, ACT: Commonwealth Scientific and Industrial Research Organisation (CSIRO) Land and Water, December 2000).

12. Study from 1999 cited in Blackmore, op. cit. note 6.

13. Need for more stringent cap from J. Whittington et al., "Ecological Sustainability of Rivers of the Murray-Darling," in *Review of the Operation of the Cap* (Canberra, ACT: Murray-Darling Basin Ministerial Council, 2000) and author's communication with scientists at *Riversymposium 2001*, Brisbane, Australia, 27–31 August 2001.

14. The focus of the Living Murray program initially is on six specific ecological assets, as described in MDBC, "Implementing the Living Murray, at thelivingmurray.mdbc.gov.au/implementing, viewed 11 April 2005.

15. Kelly, op. cit. note 6.

16. Edwards Aquifer Authority Web site, at www.edwardsaquifer.org, viewed 12 April 2005.

17. Kelly, op. cit. note 6.

18. Jorge A. Ramirez, "SAWS, Extension Partner to Help Folks in San Antonio Conserve Water," *AgNews* (Texas A&M University), 23 January 2002.

19. Dale Whittington, "Municipal Water Pricing and Tariff Design: A Reform Agenda for South Asia," *Water Policy*, vol. 5, no. 1 (2003), pp. 61–76.

20. Richard W. Wahl, "United States," in Ariel Dinar and Ashok Subramanian, eds., *Water Pricing Experiences: An International Perspective* (Washington, DC: World Bank, 1997).

Index

Other Worldwatch Papers

On Climate Change, Energy, and Materials

169: Mainstreaming Renewable Energy in the 21st Century, 2004
160: Reading the Weathervane: Climate Policy From Rio to Johannesburg, 2002
157: Hydrogen Futures: Toward a Sustainable Energy System, 2001
151: Micropower: The Next Electrical Era, 2000
149: Paper Cuts: Recovering the Paper Landscape, 1999
144: Mind Over Matter: Recasting the Role of Materials in Our Lives, 1998
138: Rising Sun, Gathering Winds: Policies To Stabilize the Climate and Strengthen Economies, 1997

On Ecological and Human Health

165: Winged Messengers: The Decline of Birds, 2003
153: Why Poison Ourselves: A Precautionary Approach to Synthetic Chemicals, 2000
148: Nature's Cornucopia: Our Stakes in Plant Diversity, 1999
145: Safeguarding the Health of Oceans, 1999
142: Rocking the Boat: Conserving Fisheries and Protecting Jobs, 1998
141: Losing Strands in the Web of Life: Vertebrate Declines and the Conservation of Biological Diversity, 1998
140: Taking a Stand: Cultivating a New Relationship With the World's Forests, 1998

On Economics, Institutions, and Security

168: Venture Capitalism for a Tropical Forest: Cocoa in the Mata Atlântica, 2003
167: Sustainable Development for the Second World: Ukraine and the Nations in Transition, 2003
166: Purchasing Power: Harnessing Institutional Procurement for People and the Planet, 2003
164: Invoking the Spirit: Religion and Spirituality in the Quest for a Sustainable World, 2002
162: The Anatomy of Resource Wars, 2002
159: Traveling Light: New Paths for International Tourism, 2001
158: Unnatural Disasters, 2001

On Food, Water, Population, and Urbanization

171: Meat Production and Consumption: A Global Perspective, 2005
170: Liquid Assets: The Critical Need to Safeguard Freshwater Ecosytems, 2005
163: Home Grown: The Case for Local Food in a Global Market, 2002
161: Correcting Gender Myopia: Gender Equity, Women's Welfare, and the Environment, 2002
156: City Limits: Putting the Brakes on Sprawl, 2001
154: Deep Trouble: The Hidden Threat of Groundwater Pollution, 2000
150: Underfed and Overfed: The Global Epidemic of Malnutrition, 2000
147: Reinventing Cities for People and the Planet, 1999

Other Publications from the Worldwatch Institute

State of the World 2005

Worldwatch's flagship annual is used by government officials, corporate planners, journalists, development specialists, professors, students, and concerned citizens in over 120 countries. Published in more than 20 different languages, it is one of the most widely used resources for analysis. The authors of *State of the World 2005* propose that the foundations for peace and stability lie in moving away from dependence on oil, managing water conflicts, containing infectious diseases, moving toward disarmament, cultivating food security, and cooperating across borders to achieve a sustainable world.

Vital Signs 2005

Tracks major social, economic, and environmental indicators of our times.

"*Vital Signs* provides the most straightforward and reliable environmental, economic, and social information available on the planet Earth. The book delivers…facts illuminated by contexts and interconnections, often revealing causes of the problems, and pointing the way towards solutions that work."
    —*Publishers Weekly*

State of the World Library 2005

Subscribe to the State of the World Library and join thousands of decisionmakers and concerned citizens who stay current on emerging environmental issues. For 2005 the Library includes our flagship annual *State of the World 2005: Redefining Global Security, Vital Signs 2005, Inspiring Progress: Religions' Contributions to Sustainable Development*, and Worldwatch Papers *Liquid Assets: The Critical Need to Safeguard Freshwater Ecosystems*, and *Meat Production and Consumption: A Global Perspective*.

World Watch

This award-winning bimonthly magazine is internationally recognized for the clarity and comprehensiveness of its articles on global trends. Keep up-to-speed on the latest developments in population growth, climate change, species extinction, and the rise of new forms of human behavior and governance.

To make a tax-deductible contribution or to order any of Worldwatch's publications, call us toll-free at 888-544-2303 (or 570-320-2076 outside the U.S.), fax us at 570-320-2079, e-mail us at wwpub@worldwatch.org, or visit our website at www.worldwatch.org.